Electronics for Electricians and Engineers

Ian R. Sinclair

Heinemann : London

William Heinemann Ltd
10 Upper Grosvenor Street, London W1X 9PA

LONDON   MELBOURNE   JOHANNESBURG   AUCKLAND

First published 1987

© Ian R. Sinclair 1987

British Library Cataloguing in Publication Data
Sinclair, Ian R.
 Electronics for electricians and engineers.
 1. Electronics
 I. Title
 537.5′02462   TK7815
 ISBN 0 434 91837 7

Photoset by Wilmaset, Birkenhead, Wirral
Printed in Great Britain by
Redwood Burn Ltd, Trowbridge

Contents

Preface	vii
1 Fundamentals	1
2 Electrostatics	11
3 Electric current	25
4 DC circuits	37
5 Electricity and heat	56
6 Magnetism	67
7 Current and motion	82
8 Ions	103
9 AC circuits	113
10 Electrons in a vacuum	135
11 The semiconductor diode	146
12 Transistors	168
13 Linear ICs	201
14 Digital integrated circuits	220
15 Practical matters	252
Index	265

Preface

The rapidly changing technology of electronics has left many technicians requiring an urgent updating of their skills. The same problems have faced many in other fields of engineering who find now that they require a knowledge of electronics in order to understand new developments in their own subjects. Hardly any part of modern life is now untouched by electronics, and there can be very few people working in any branch of engineering or science who will not find that some understanding of electronics is necessary.

This book has been written in response to the need that now undoubtedly exists. The aim has been to explain principles and devices in clear terms, assuming no high level of prior knowledge. I have assumed only that the reader will have some elementary knowledge of electricity, more from a practical than a theoretical standpoint. For that reason, the early chapters of the book are concerned with a review of modern electrical principles, and may be omitted by anyone who is thoroughly familiar with them. For the reader whose theoretical knowledge may have been dulled by time, these chapters should provide a very useful revision of topics that are essential to the understanding of electronics. In the chapters that deal with electronics, the emphasis has been on principles and devices, but sufficient circuit diagrams have been included to illustrate how the various electronic components are used.

I am most grateful to RS Components Ltd for considerable help and support in this project, particularly with the provision of photographs and datasheets on modern electronic devices.

Ian R. Sinclair

1
Fundamentals

Asking what electricity *is*, as distinct from how electricity behaves, is not always a help to understanding the behaviour of conventional electrical circuits. When we need to understand how *electronics* devices work, however, some idea of what electricity is becomes essential. In this book, we are concerned not with the kind of precise detail that you might find in a textbook of physics, but with the essentials only. These essentials start with the idea of **atoms**.

We have known for almost two hundred years that all the substances we know on earth and can detect anywhere else in the universe, are built up from about one hundred basic materials that we call **elements** – substances like carbon and silicon, oxygen and hydrogen. Each element is made up from huge quantities of identical units called atoms, one type of atom for each element. This idea, first put forward by the ancient Greeks and in its present form by John Dalton, was the foundation of the study of chemistry in its modern form. It was originally thought that the atoms of a material were indestructible and unchanging, but as evidence accumulated, it became clear by the end of the nineteenth century that this was not so. Atoms seemed in turn to be made up from smaller particles; so much smaller in fact, that each atom was mainly empty space in which tiny particles moved like the planets round the sun in our solar system. This picture of the atom is one that has been very useful, though for some purposes it is inadequate.

If we stay with the 'solar system' idea of the atom, then the place of the sun in this picture is occupied by the **nucleus** of the atom, and the planets by the **electrons**. As far as we are concerned, the central core of the atom, the nucleus, is not of great importance, though it is the part of the atom that is of most interest to a chemist or, of course, a nuclear physicist. The tiny particles, or electrons, that surround this nucleus are responsible for electricity and electronics. We don't know much about what these particles are, except for suspicions that they might be made up out of even smaller particles. We *do* however, know a lot about what they do, and that's the foundation of all electrical science.

The most important feature of electrons is the force that keeps them in place in the atom. This force is called the *electric* or *electrostatic* force; it attracts the electrons to the nucleus and repels them from each other. The strong electric force is balanced by other forces, and this balance of forces ensures that the electrons are held in the atom, but without collapsing completely into the nucleus. It is the balance of forces that ensures that the atom is mainly empty space with a cloud of electrons surrounding the nucleus. Note that the electrons are so small that they can easily pass through the mainly empty space of atoms.

Before we knew about electrons, the effects of the force were known, and were put down to something called **charge**. We still don't know what charge is, but we do know that the electron is the source of it. For historical reasons, we talk of charges having two signs, positive and negative, and the type of charge on the electron is negative. The charge type of the nucleus is positive, and the unit amount of charge is the amount on the electron. We have never found any amount of charge less than this, so there are no fractions of this amount. The idea of charge is useful, because it allows us to form an idea of the way that the nucleus and the electrons are arranged.

In the complete atom, the amount of charge is normally zero. This is because each electron in the atom has a unit negative charge, and the nucleus has a positive charge whose size is equal to the sum of the charges on the electrons. If, for example, an atom consists of a nucleus with 12 electrons, then the nucleus will have 12 units of positive charge, making the atom as a whole neutral. Any substance that has positive charge must have had electrons removed, and a substance that has negative charge must have gained some electrons. This is because the electrons are the only part of the atom that can be comparatively easily removed, temporarily at least.

The balance of forces in the neutral atom ensures that when an electron is removed from or added to the atom, there will be strong forces trying to reverse the situation. These forces are of measurable size when more than a few electrons are involved, and it was from measurements of these forces that we first gained some knowledge of electricity. Even in ancient times, it was known that similar charges repelled each other, and opposite charges attracted. The more precise laws were discovered by Coulomb, forming what we now know as **Coulomb's Law**.

In these days, substances could be classed as **insulators**, which retained charge, and **conductors**, which would lose it. We know rather more about materials now, but the general principles are unchanged. An insulator is a material whose atoms are arranged in such a way as to prevent electrons from moving easily. A spare electron on the surface of an insulator will therefore stay where it lands. A conductor is made from atoms whose arrangement allows electron movement from one atom to another. An electron on the surface of a conductor can easily move to any other part, and if there happens to be any part that has lost an electron, then the movement will ensure that the lost electron will be replaced in a very short time. This essential difference between insulators and conductors is of vital importance both in electricity and in electronics.

Electrostatics and current electricity

When electrical effects were first discovered, it was thought that there were two types of electricity, static and moving. We now know that the differences are due to the way in which electrons were being used. If we shift some electrons to or from an insulator, forces can be measured between this insulator and anything around it. Only small numbers of electrons are involved, so that the total amount of charge is also very small. This is because the forces are so very large that trying to move more than a few electrons is almost impossible – the electrons will return through the air, or even through a vacuum, in what we call a spark. This behaviour is what we now call **electrostatics**. When we allow electrons to move through conductors, by contrast, very large numbers of electrons can be moved because the movement is one of shuffling from atom to atom, with no atom ever losing or gaining an electron for any noticeable time. The forces on the electrons and so on the materials are very small even though such large amounts of charge are being shifted. This behaviour is what we call **electric current**.

The important distinction is that electric current moves in a closed path, called a 'circuit', but electrons on an insulator are held at rest. Very large forces act on these electrons at rest, but not on the moving electrons because in a circuit there is no atom which has a surplus or deficit of electrons for more than a very short time. This does not mean that there are no forces acting on the electrons in a circuit. The forces are very different, however, and are caused by the effect of the movement of the charge rather than by the charge itself. We call these types of forces **magnetism**.

Both the electrostatic forces and the magnetic forces have an effect at a distance, unlike mechanical forces. The electrostatic forces operate on anything that carries an electric charge at rest, with the usual direction – attractive for charges of the same sign and repulsive for opposite signs of charge. The magnetic effects operate on any moving charge and on any material that contains an unbalanced moving charge. This includes both electrical circuits and certain materials such as iron. Just as we can use the amount of electrostatic force to measure the amount of static electrical charge, we can use the amount of magnetic force to measure the amount of moving charge. There is an important difference, however. When we work with electrostatics, the amount of electrons that we can ever displace is very small, and the total force between two charged materials is small. When we work with electrical currents, we are dealing with unbelievably large numbers of electrons, and though the magnetic force for each electron is very small, the huge numbers cause this to add up to a very substantial force. We therefore use this force effect in measuring instruments such as ammeters and voltmeters.

So far, we have thought of the atom in terms of nucleus and electrons, with electrons being shifted to or from the atom. In fact, we can normally only add or remove one electron per atom, because the forces between the electron and the rest of the atom are so strong. When one electron is removed or added, the resulting atom will have an electric charge, and will behave very differently from an ordinary atom. It has become an **ion**, and will try to become neutral again by adding or shedding an electron. We can very seldom strip all the electrons from a nucleus, so that the way that a nucleus behaves is of no great interest as far as electricity and electronics is concerned. The behaviour of ions, by contrast, can be quite important.

Ions can exist in gases and in liquids. When the atoms of a gas become ions, the gas which is normally an insulator can conduct electricity. Similarly when a liquid contains ions, the normally insulating liquid will conduct electricity. The side-effects of the existence of ions are even more interesting and useful. The ions of a gas, moving because of the electrostatic forces on them, will collide with each other and with electrons. When the ions revert to neutral atoms again, the energy that they absorbed in order to become ions is given out again, but in the form of light. Ionised gas tubes are therefore used as neon signs, decorative lights, and in fluorescent lighting. Most of the ionised liquids that we use are solutions in water, and when the ions move in such solutions, there are chemical changes when the ions give up their charges. These allow such effects as electroplating, the isolation of elements that do not occur naturally (like aluminium), and electrolytic polishing.

Conductors and insulators

The old classification of materials into conductors and insulators was rough and ready, but it served for a century or more. Nowadays, we need to be more specific about materials, not least because we know so much more about the movement of electrons. Basically, the difference between a typical insulator and a typical conductor is in how the atoms of the material are packed together. In an insulator, each atom is fairly isolated. An electron lost by or gained by an atom does not cause an electron to move to or from the next atom because distances are too great (by electron standards). In a conductor, by contrast, atoms are packed tightly together, and the nucleus of one atom can even affect the electrons in the next atom. In addition, each atom is of a type that contains a large number of electrons that are some considerable distance from the nucleus and so less strongly bound to the nucleus. This means that the material is dense, strong, and all the other things that we associate with a metal – most common conductors are metals, and all metals are conductors.

The obvious difference between conductors and insulators is that of **resistivity**, the quantity that measures resistance of a material to the flow of electrons. Conductors allow electrons to flow easily through them, and they even allow other movements of charged objects, the objects that we call 'holes'. There are other differences, however. One important difference concerns the effect of raising the temperature of a material. When you raise the temperature of a conductor, the flow of current becomes less easy. More precisely, the resistivity of the material increases (see Figure 1.1). This is because heating the material increases the vibration of all the particles in the material, and that in turn makes it more difficult for electrons to thread their way through the atoms. By contrast, when you heat an insulator you make it easier for electrons to move through it. The vibration of the atoms in the hot material makes it likely that some electrons can break free and move, even if this movement is limited.

There is yet another important difference. Suppose you take two elements, one metal and one non-metal. Elements, in the strict chemical sense, means that each material is made out of its own type of atom, with no other atoms present. Nothing is ever so perfectly pure, but we can prepare many types of elements now in which only one atom in a thousand million (or more) is an impurity atom. Given two pure materials like this at normal temperatures, the metal will be a conductor, and the non-metal will be an insulator. Now the effect of impurity on such materials will not be very dramatic. Adding atoms of a different metal to a metal element will not greatly affect its electrical resistivity unless the addition is on a large scale, certainly 1 per cent or more, and even then the effect is not large. Similarly, adding another non-metal to the non-metal element does not very noticeably affect its resistivity. The elements that are good conductors or good insulators do not have these characteristics greatly changed by the presence of impurities. It is just as well, because we would know rather less about electricity if this had not been the case.

Semiconductors

Semiconductors are not simply materials whose resistivity (or its inverse, conductivity) is somewhere between that of a conductor and that of an insulator. Certainly, one feature of a pure

> The *resistance* of a sample of material depends on the dimensions of the sample and on the material from which the sample is made.
>
> *Resistivity* is a factor that measures the effect of a material on the resistance of any sample. The dimensions of a material are affected by temperature changes, but the effect of temperature on resistivity is very much greater.
>
> As a formula:
>
> $$R = \frac{\rho \cdot s}{A}$$
>
> where ρ is the resistivity (units:ohm-metres), s is the length (units:metres) and A is the area of cross-section (units:metres squared).

Figure 1.1 *Resistance and resistivity. Resistance depends on temperature mainly because resistivity depends on temperature, though the dimensions of a resistor are also slightly affected by temperature*

semiconductor is that it will have a resistivity value that is not so high as that of an insulator, but it certainly does not approach the value that we would expect of a conductor. The two features that make us class a material as a semiconductor are the effect of temperature and the effect of impurity, and both of these effects are closely related.

Suppose, for example, that we have a specimen of pure silicon. Its resistivity is very high, in the same range as most insulators, so that we would normally think of this material as an insulator. When the pure silicon is heated, however, its resistivity drops enormously. Though the drop is not enough to place hot silicon among the ranks of good conductors, the contrast with any other insulators is quite astonishing. Even more remarkable is the effect of impurities. Even traces of some impurities, one part per hundred million or so, will drastically change the resistivity of the material. It is because of this remarkable effect of impurity that we took so long to discover semiconductors – it was only in this century that we discovered methods of purifying elements like silicon to the extent that we could measure the resistivity of the pure material. It is interesting to note, incidentally, that a lot of this research was done during the great depression of the 1930s, and many politicians thought at the time that all this useless research should be scrapped and the money spent on something useful, like the dole.

The main difference between a semiconductor element and any other element is that a semiconductor can have almost any value of resistivity that you like to give it. It is, in other words, a material that can be engineered to have the characteristics that you want of it. The manipulation is done by adding very small quantities of other elements. These cannot be just any elements, however. Semiconductor elements, like metal and many non-metals, form crystals. A crystal is the visible evidence of the arrangement of atoms, and many materials have atoms that will arrange themselves into patterns because of the forces that exist between the atoms. When you add an impurity to a crystalline material, the atoms of the impurity have to take a place in the crystal. If the impurity atoms are very different in size, the result will be to distort the crystals, but it is usually possible to find elements whose atoms are about the same size as the atoms of the semiconductor material. When such atoms are added, it is likely that they will fit neatly into the crystal, taking the place of the normal atoms of the semiconductor. That is one requirement: fitting into the crystal. The other requirement is one of structure. The impurity atoms must not have the same arrangement of electrons as those of the semiconductor.

It's easier to see what is required if we look at some definite example. Pure silicon has atoms in which there are 14 electrons, but of these electrons, four are much less strongly bound – we can think of them as an outer layer. Now there are two elements, boron and phosphorus, that have atoms of fairly similar size, close enough to fit into the crystals of silicon. Of these, boron has only three electrons in its outer layer, and phosphorus has five. Both of these materials greatly affect the resistivity of the pure silicon, because they affect the availability of electrons. The phosphorus atom has one outer electron more than the silicon atom, and this will be set loose when the phosphorus atom is fitted into the crystal. The boron atom has one electron less in its outer layer, and this atom will trap an electron from a silicon atom. Either impurity causes a massive drop in resistivity, so that the material becomes a conductor.

Electrons and holes

Even in the nineteenth century there was a suspicion that electric current through solids was not all caused by electrons. Due in a very large extent to the work of a physicist called Hall, we discovered that there are two ways that electric current can be carried in crystals (note that this applies only to crystals). Crystals are never perfect, and when a crystal of an almost pure material has been deliberately made impure (or *doped*), the crystals contain atoms of a different type. If these atoms possess more or fewer electrons in their outer layer than the normal atoms of the crystal, then the electrical characteristics will also change. One way of changing the characteristics is to release more electrons. The other way is to release more holes. A hole is a part of a crystal that lacks an electron. Because of the structure of the crystal, a hole will move from atom to atom and when it does, it behaves just as if it were a particle with a positive charge. Within the crystal, the hole has a real existence, we can measure its charge and even a figure for mass. The important difference is that the hole is a discontinuity in a crystal, it has no existence outside the crystal. The electron, by contrast, can be separated from the crystal, and can even move in a vacuum (see Chapter 10).

Most metals exist as crystals, and are by no means very pure. As a result, holes exist in the crystals, and contribute to the flow of electric current. In a few metals (one is zinc) more of the current may be conducted by the holes in the crystals than by the electrons. Hole conductivity is even more important in semiconductors because we can control it. As we have described, doping a pure silicon crystal with boron will create holes, and make this silicon conduct mainly by hole movement. Silicon doped in this way is called **p–type**, the p meaning positive. This doesn't mean that the crystal has a positive charge, only that most of the current that flows through it will be carried by the positively charged holes. If, by contrast, we dope the pure silicon with phosphorus, the extra electrons released in this way ensure that most of the current flow is because of moving electrons. Because the electron is negatively charged, we call this doped silicon **n–type**. Once again, this does not mean that the material is negatively charged, only that most of the moving charged particles are electrons. Doped semiconductors, like all other solids, are electrically neutral; for each positive charge there is a negative one. The difference between p–type and n–type is decided by the charges that can move as distinct from the ones that are tightly bound into the atoms. Not only can we make a semiconductor have the amount of resistivity that we want (within limits), but we can decide which type of conductivity it will be. We will come back to the importance of all this in Chapter 11, when we look at the semiconductor diode and how it works.

2
Electrostatics

Electrostatics is about electric charge at rest, which in turn means electrons transferred from one place to another and staying in place. This transfer is not difficult: rub one insulator against another and a large number of electrons will be transferred, making each insulator charged. The presence of the charge will cause the insulators to be attracted to each other, and you can sometimes detect a spark if they are placed close to each other. The spark is the visible sign that the electrons have returned to their places through the air. Most of the effects of what we now call electrostatics had been noted by the middle of the eighteenth century; this was the first form of electricity to be discovered.

Charge and potential

Though the forces between charged insulators are certainly measurable, and we can measure amounts of charge by way of these forces, it is certainly not easy to measure charge in this way. The reason is that the amount of force between two charged objects depends on the distance between the charges, and it is not easy to know exactly where the charges are located. Experimenters like Coulomb used charged objects that were metal spheres supported on insulators, and assumed that the charge was concentrated at the centre of each sphere. This is a reasonable approximation, but since the charge on a metal can move about, it is not ideal. Small inaccuracies like this in the measurement of distance have a large effect on the result when we try to measure charge by the amount of force that it exerts. That's because the distance in the formula is squared (see Figure 2.1), which has the effect of magnifying any errors in distance measurement. It is quite remarkable, in fact, that Coulomb obtained results as precise as he did.

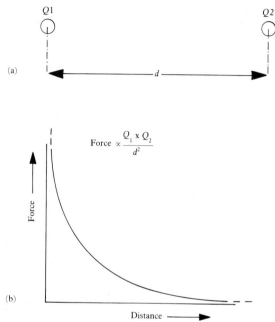

Figure 2.1 *Electrostatic force and distance. (a) The relationship between force, charge (Q) and distance (d) shows that the amount of force between charges depends on the product of the charge amounts divided by the square of the amount of distance. The graph (b) shows how force becomes negligible at larger distances, typically a few centimetres*

There is another effect of electric charge which is much easier to measure. When charge has been moved, some mechanical work is needed to move it. This amount of work divided by the amount of charge gives a quantity called **potential**, and the unit of measurement is called the **volt**. Modern instruments allow us to measure potential much more precisely than we could measure the mechanical force between charges, or the amount of charge, so that potential is a much more useful quantity to work with. In particular, the difference in potential between two points is easily measured, and is known as the PD (potential difference) or more usually, the **voltage**. From this measurable quantity, we can calculate quantities that are much less easy to measure. If, for example (Figure 2.2), we have two points with a potential difference of V volts between them, and a distance of d metres apart, then the quantity V/d is called the electric field strength, and its units are volts per metre. The force on a charge of size Q coulombs placed on the line between these points will then be Q × electric field strength, which is QV/d. This gives us another method of measuring the amount of charge in terms of force, distance and voltage, all easily measurable quantities.

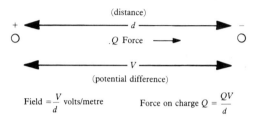

Figure 2.2 *Potential difference, distance and electric field*

A high potential difference exists between two points when a lot of mechanical work has to be done in moving a small electric charge from one point to another. This is done when two insulators are rubbed together, and it can be done by methods that are purely electrical, as we shall see in Chapter 11. The old Wimshurst machine and the more modern Van der Graaf generator are purely electrostatic methods of creating high potential differences, and are still used to demonstrate the principles, though they have no place in this book.

Electrostatics deals with small amounts of charge which are produced when electrons separate from atoms, and remain separated. Even the separation of a very small number of electrons can cause very high potential differences, in the order of thousands of volts, but the effect of these large potential differences is very small because so little charge is involved. Touching a small sphere which is at a potential of several hundred thousand volts produces no more than a mild tingle, because the effect of electricity on the body depends on the current that flows (the amount of charge per second) rather than on the potential difference. The voltmeters that we use for measurements on conventional electric circuits are of no use for measuring electrostatic voltages, because such voltmeters depend on passing current, and charged objects don't have enough charge to pass much current.

Electrostatic voltmeters in the past relied on the forces that are produced when a large potential difference exists, and some could detect voltages as low as 300 V. Nowadays, it is more common to use electronic voltmeters with adaptors (in the form of resistors) to measure the high voltages that are used in some electrostatic equipment. Since all of the electrostatic equipment that you are likely to use makes use of mains-powered generators, higher currents can be passed, and comparatively conventional voltmeters can be used. You should remember, however, that this means that the high voltages will be dangerous, much more dangerous than the voltages generated by a purely electrostatic generator such as a small Van de Graaf generator.

The electric field

The existence of a potential difference between two points means that there is an electric field between these points. The size of the electric field is equal to the potential difference between the points divided by the distance between the points, and its units are volts/metre. For example, if you have two points 10 cm apart and with 50 000 V between them, then the field strength is 50 000/0.1 = 500 000 V/m, or 500 kV/m. In any electric field, there will be a force on any charged object, and even uncharged objects will be affected. If you think of a piece of hair, for example (Figure 2.3), the effect of an electric field will be to separate charges in the material, making one end positive and the other negative. The hair is not charged, because the positive and negative charges are equal, but the positions of the charges are not the same. As a result, the hair will turn to line up with the field just as a compass needle lines up with a magnetic field. We usually represent an electric field on a diagram with arrows that indicate the direction in which threads or hairs will line up. For historic reasons, we always show the arrows pointing to the negative end of the field, so that the direction of the field is taken as being from positive to negative.

A charged object will not simply turn in an electric field, it will move along the line of the field. An electron, for example, will

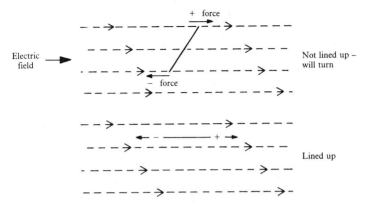

Figure 2.3 *The effect of an electric field on long strips of insulating materials is to turn them into line with the field*

move towards the positive end of the field, and a positively charged particle such as a positive ion, will move towards the negative end of the field. Electrons and ions by themselves are invisible, but they are usually part of other particles. For example, the air consists of gas atoms which can become ionised in electric fields, and then move. Particles of dust are nearly always charged, either by rubbing against each other, by movement through the air, or by the effect of an electric field. Any electric field can therefore be expected to cause dust to be attracted. The high voltage on the inside of the screen of a TV receiver, for example, is modest by electrostatic standards, only around 15–22 kV. This also appears as an electrostatic voltage on the outside of the glass, and will attract dust, as you can see after only a few days of use. This feature of an electric field is used commercially in dust precipitators.

Using electrostatics

The three main applications of electrostatic fields, apart from devices like TV tubes (see Chapter 10), are in air cleaners, electrostatic loudspeakers, and ionisers. Electrostatic air cleaners are used extensively in air-conditioning systems, and operate by passing the air through metal meshes (Figure 2.4). These meshes are held in insulating frames, and are connected to a supply so that there is a large potential difference (typically 20–50 kV)

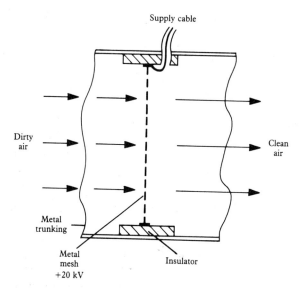

Figure 2.4 *The principle of electrostatic precipitators for cleaning air*

between the mesh and the other metal in the air trunking. If, for example, the mesh is at a positive potential, it will attract any particles of dust that have acquired a negative potential, and if the mesh is at a negative potential, it will attract any particles that are positively charged. As it happens, most dust particles are negatively charged, and by using a mesh at a positive potential, these particles can be attracted towards the mesh. The great advantage of electrostatic air cleaning as compared to the use of filters is in the size of particles that can be removed. Conventional air filters deal well with the usual dust from fabrics, but the air has to be pumped through such filters at a fairly high pressure in order to maintain a good flow of air. Electrostatic precipitators will remove even very small particles of smoke from the air, and are particularly useful for cinemas and hospitals. The mesh can be of much more open structure than a conventional filter, allowing air to circulate more easily with lower powered fans and so less noise. It is even possible to monitor how much dust is being collected by measuring the current flowing to or from the mesh, since each dust particle striking the mesh will neutralise some charge that will then have to be replaced by the power supply. This current is very small, a microamp or less, but is measurable by electronic instruments.

The use of electrostatic precipitators, TV receivers, and any other equipment that makes use of high voltages, will create ions in the air. These ions can be negative or positive, according to whether the electric fields remove electrons from the atoms or donate them. There is a widespread belief, not entirely supported by medical evidence, that ionisation of the air affects the well-being of anyone breathing the air. A few air-conditioning systems that use electrostatic precipitators employ two meshes, one positive and one negative, to ensure that the delivered air is free of ions. Several manufacturers now supply air ionisers for both public buildings and domestic purposes, and these operate along the lines of precipitators, using a fan to move the air through a charged metal mesh whose potential will ensure that ions of the unwanted polarity are removed.

Electrostatic loudspeakers use the forces on charged objects in an electric field to move a large diaphragm. A conventional (electromagnetic) loudspeaker uses a comparatively small cone of stiff material which is moved in and out by magnetic action. To reproduce sound at realistic levels, and in particular to reproduce the lower notes, the cone of a conventional loudspeaker has to be moved over comparatively large distances, or many loudspeakers have to be connected together. In the full-range electrostatic loudspeaker (Figure 2.5), a large light plastic diaphragm is stretched between two metal meshes. The plastic is metallised on its surfaces, and is connected through a high resistance to a 'polarising' voltage of around 6 kV. The meshes are connected to the audio signals, with one signal inverted with respect to the other. As the voltages on the meshes rise and fall, the forces on the diaphragm in the electric field will make it move to and fro, reproducing a sound wave. Wide range electrostatic loudspeakers, particularly those manufactured by the Acoustical Company under the QUAD brand-name, have earned a world-wide reputation for excellent sound reproduction.

Varying electric fields

Electrostatics is about electric fields that do not change because charges are not being moved. We can't leave the topic, however,

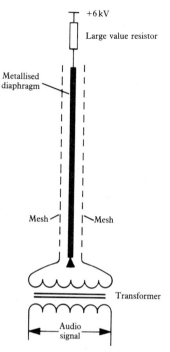

Figure 2.5 *The principle of the full-range electrostatic loudspeaker. The audio signal comes via a transformer which greatly increases the voltage amplitude and ensures that one signal is inverted*

without some mention of what happens when an electric field changes, because this is of fundamental importance in electronics. The effect of a varying electric field was predicted by Clark Maxwell in 1862 and confirmed by Heinrich Hertz in 1887. When an electric field is changed, either suddenly, as in a spark discharge, or in the form of an oscillation, then electromagnetic waves are generated. A changing electric field causes a magnetic field to be produced. This magnetic field is also a changing field, and it generates another electric field, also varying. The effect of this continuing regeneration of the effects is a wave of electric and magnetic fields moving outwards from the original source, and carrying energy. This is the type of wave that we call a radio wave, but the general term is **electromagnetic wave**. A radio wave is simply an electromagnetic wave of one particular range of wavelengths, from a few millimetres to many kilometres distance between wavepeaks. The same type of wave motion is also described as light, radiated heat, infra-red, ultra-violet, gamma radiation, and so on, depending on its wavelength. In other words, if you want to live in a world free from the effects of radiation, you must first turn off the sun!

Capacitance and capacitors

Apart from the devices like cathode-ray tubes that make obvious use of electric fields, the main impact of electrostatics on electronics is through the use of **Capacitors**. A capacitor is any device that can store charge, and by this definition, almost anything can act as a capacitor. A metal sphere carried on an insulated holder is the simplest form of capacitor, and if we look at what happens in such a sphere as charge is taken up to it, we can understand much more easily what happens inside capacitors of more conventional construction.

To start with, imagine the sphere uncharged. When some charge is brought to the sphere, mechanical work has to be done, and the sphere gains a potential. The amount of the potential is just the amount of work that had to be done shifting the charge, divided by the amount of charge. The more charge that is brought up to the sphere, the more the potential increases. As all this happens, however, the ratio of the total charge on the sphere to the potential of the sphere remains constant (Figure 2.6). This quantity, charge/potential, is called **capacitance**. Its unit is a coulomb/volt, but we normally use the alternative name, **farad**.

For a sphere of fixed dimensions, the capacitance is a fixed amount and is very small. Larger spheres have larger amounts of capacitance, and this was an observation that led people to believe once that electricity was some kind of liquid. The capacitance of anything depends on its dimensions and on the arrangement of conductors, and we have devised more convenient arrangements than the sphere. In particular, we make use of the potential difference between conducting plates, an arrangement called the 'parallel-plate' capacitor.

The basic arrangement of the parallel-plate capacitor is shown in Figure 2.7. The two conducting plates are separated by an insulator which can be air, but which is more likely to be some solid insulating material called the **dielectric**. The capacitance

20 Electronics for Electricians and Engineers

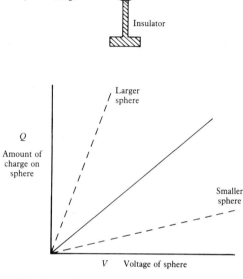

Figure 2.6 *Charging a sphere. The ratio of charge to voltage is constant for any one sphere, and is called the capacitance of the sphere*

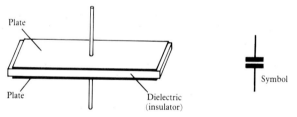

Figure 2.7 *The arrangement of a parallel-plate capacitor. This is the fundamental type of capacitor used in electronics*

mainly exists between the plates, not from one plate to the surroundings, so that we can control the size of the capacitance more easily. The amount of capacitance depends on the dimensions of the arrangement, and on the insulating material. As you might expect, the capacitance depends on the area of one of the two (assumed identical) plates. The larger the plate area, the greater the capacitance value. The capacitance also depends on the spacing between the plates. In this case, however, the

Electrostatics   21

smaller the distance between the plates, the greater is the capacitance. This is because a smaller spacing between the plates increases the field between the plates. In addition, the insulator contributes a factor, called **relative permittivity**.

Figure 2.8 shows the formula for a parallel-plate capacitor. The dimensions are in metres of distance, and square metres of area, but the relative permittivity has no dimensions, it is just a number. This number is 1 for air, and from 2 to 10 for the type of insulators that are normally used. The capacitance value that is found from this formula is in farads. Since the farad is a very large unit, we usually quote capacitances in microfarads (millionths of a farad), nanofarads (thousandths of millionths of a farad) or picofarads (millionths of millionths of a farad).

$$C = \frac{\varepsilon_o \varepsilon_r A}{d}$$

A = area of one plate (assumed equal)
d = distance between plates
ε_o = permittivity of free space
ε_r = relative permittivity of dielectric

Units: A in m^2
d in m
ε_o = 8.85 x 10^{-12}
C in farads

or

A in mm^2
d in mm
ε_r = 8.85
C in nF

Figure 2.8 *The parallel-plate capacitor formula. Capacitance is increased by using large plate area and by spacing the plates very close together*

The formula itself is not of particular importance unless you happen to be in the business of manufacturing capacitors, but the meaning of it is important. To create a large capacitance, you need plates of a large area spaced very close to each other and separated by a dielectric with a large value of permittivity. The formula is also of importance if you are trying to avoid unwanted (stray) capacitance. You will get unwanted capacitance problems if you have large areas of conductors at different voltages close to each other, particularly when there is a solid insulator between them. To avoid such 'stray' capacitance, designers make use of bare wires whose paths cross at right angles and are separated as widely as possible. You will see these techniques used in equipment in which stray capacitance must be kept down to a minimum.

The capacitors that are used for electronics circuits are all based on the parallel-plate design. Small capacitance values make use of thin plates of mica or ceramic that are coated on each side with metal layers. Rather than make the plates bigger, larger capacitance values can be obtained by stacking plates (Figure 2.9), connecting alternate layers together. This type of construction is used for values of a few picofarads (pF) up to several nanofarads (nF). For large values, the rolled construction (Figure 2.10) is used. The capacitor consists of a long thin strip of insulator coated on each side with metal. In the past, oiled paper has been used as the insulator, and the metal was in the form of strips of foil rather than a coating on the paper. Nowadays, the insulator is more likely to be some form of plastics material, such as polyester, polystyrene, polycarbonate, polypropylene or others. The insulator and its metallised coating is then rolled up tightly with another strip of non-metallised insulator. In this way, large surface areas of 'plates' (the metallising) can be obtained in a capacitor of quite small overall size; because very thin films of plastics materials can be manufactured, the spacing between the plates is small and capacitance values can be large.

The ultimate in large capacitance value is provided by

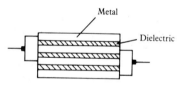

Figure 2.9 *Connecting alternate plates in a pile of parallel plates*

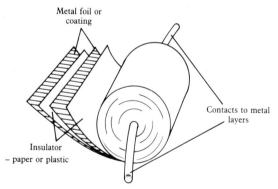

Figure 2.10 *The rolled construction for a capacitor*

'electrolytic' capacitors. These use aluminium or tantalum foil which has been coated with a very thin film of metal oxide. The coating is done by immersing the metal in a mild acid, often in the form of a jelly, and applying a voltage between the foil and the outside casing. The voltage that is applied in this way is called the 'polarising voltage', and once this has been applied, any voltage applied to the capacitor must be in the same direction – foil positive, case negative. Because the foil can be etched and dimpled to present a very large area, and the insulator is the thin layer of oxide over the foil, the capacitance values can be very large, up to a farad or even more in a compact unit. The thin layer of oxide is fragile, however, and so the voltage that can be applied across an electrolytic capacitor is limited – it may be as low as 3 V, and is seldom more than 500 V. The value of capacitance is also rather variable – a manufacturing line for 16 μF capacitors may turn out values that range from 14 μF to 24 μF. Modern electrolytics can be obtained with much tighter tolerance of value than was formerly possible, typically 20 per cent either way, and solid-electrolyte capacitors can be obtained which permit the voltage across the terminals to be reversed. All electrolytic capacitors, however, suffer from leakage current; they will lose their stored charge at a rate that depends on temperature and capacitance, typically 10–500 μA of current. This leakage, however, still allows very large value capacitors to be used in place of batteries – the RS Components catalogue lists a 3.3 F capacitor with leakage of 820 μA at rated voltage, offering retention of charge for several hours. Such capacitors are used in computer memory boards to provide a back-up in case of power supply failure, since capacitor back-up is more reliable than battery back-up for many purposes.

The amount of charge that a capacitor stores depends on its capacitance value and on the potential difference across its terminals. The formula for charge stored is $Q = CV$, meaning that the charge (in coulombs) is equal to the capacitance (in farads) multiplied by the voltage (in volts). Taking the example of the memory back-up capacitor, the charge on a 3.3 F capacitor with a potential difference of 3 V is $3 \times 3.3 = 9.9$ coulombs. If you are working with capacitance values in microfarads, then the charge values are in microcoulombs, and using capacitance units of nF and pF give charge units of nC and pC respectively. These amounts of charge become important when we are working with capacitors in pulse circuits, where capacitors are being charged and discharged through or across a resistor.

3
Electric current

When electric current was first observed, using batteries, no-one suspected any connection between current and electric charge. This is because electric current flows in circuits, using low potential differences, and with none of the effects that we see when charge is permanently moved from one place to another. We know now that electric current is just the flow of charge, and is defined as *the total amount of charge passing any point in a circuit per second*. Since current flows in a closed path, the current can be measured at any point in the same circuit. If no charge is permanently shifted, then the current at any point in a circuit should be the same as the current at any other point at the same instant of time. The units of current are coulombs per second, but the more common name for this is the **ampere**, usually abbreviated to amp (symbol A).

Currents in early circuits were steady because the voltage of a battery (its electromotive force, or EMF) is steady. Steady currents are called DC, direct current, and only later was it discovered that a different pattern of current could be generated. The first generators that used rotating coils and a magnet gave AC, alternating current. The name is used because the direction of the current alternates, changing direction at regular intervals. When AC is used, the terms positive and negative become less important in a circuit, because voltages in AC circuits, like the currents, change direction at regular intervals. Electronics brought a whole new type of currents and voltages in the form of pulses. A pulse (Figure 3.1) is a sudden change, followed by a slower return to normal. If, for example, the current in a cable is

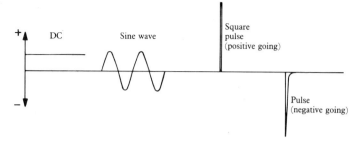

Figure 3.1 *DC, sinewave and pulse voltages. A pulse can be a brief square wave with both 'sides' almost vertical, or the more usual type in which only the first edge, the leading edge, is vertical. The pulse can be positive-going or negative-going*

zero, then suddenly rises to 1 A in a time of one microsecond (one millionth of a second), and then falls back to zero over a time of 10 µs, then this is a pulse of current. The current in this example is unidirectional (it flows in one direction only, unlike AC), but is not steady. In this and the next few chapters, we shall be dealing with the effects of direct current only.

Direct current

Electric current, of the DC variety, flows in a circuit which is a closed path made of conducting material. The best conductor of all is silver, but because of the high price of silver, copper is the normal material for constructing conductors. Aluminium is also used for some purposes, notably high-voltage supply cables that pass comparatively low currents.

In order for current to flow in a circuit, a voltage supply must exist at some point in the circuit. The old name for a voltage supply is *a source of EMF* (electromotive force), and originally this would have been a **cell** or battery of cells. A cell is a chemical device that converts the energy of a chemical reaction, such as a metal dissolving in acid, into electrical energy. In a cell, electrons are forced to move from one metal to another, making the metal that loses electrons become positive, and the other metal negative. The number of electrons that is needed to establish a potential difference in this way is very small, smaller than would occur in an electrostatic generator. Most chemical cells come to a balance of electron movement when the potential difference between the metals is about 1.5 V, a long way from the tens of thousands of volts that we can generate electrostatically. When the cell is connected as part of a circuit, however, huge numbers of electrons can be shifted because there is still very little permanent displacement. The chemical energy provides the effort that is needed to keep the movement going.

A DC circuit, then, operates with low potential differences and comparatively high currents. There has to be a source of EMF for current to flow, and there must be a closed circuit. Without the circuit, there is no current, and without the EMF there is no current. Cells are still a useful source of EMF for a number of purposes, particularly for portable equipment, and many types have been evolved over the years. One of the most enduring is the carbon-zinc cell, illustrated in Figure 3.2. This uses a centre rod of carbon, which will become the positive terminal. This in turn is surrounded by chemicals in jelly form, one of which is a mild acid, and the outer case of the cell is zinc. The acid dissolves the zinc, transferring electrons to the zinc so that the zinc case becomes negative, leaving the carbon rod positive. The action is very slow when the cell is not connected, so that the cell has a fairly long shelf-life. The cell fails either when the zinc dissolves completely (leaving the acid to eat its way through your cycle

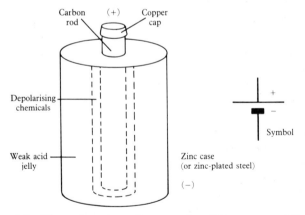

Figure 3.2 *The carbon-zinc cell construction. This uses a mildly acid material, ammonium chloride, in jelly form to dissolve zinc and so generate an EMF. The carbon rod is unaffected by the acid and is the positive pole. The depolarising chemicals remove hydrogen gas bubbles which otherwise coat the carbon with an insulating layer*

lamp or radio) or when the other chemicals are exhausted. One of these other chemicals is known as the depolariser; its job is to remove traces of hydrogen gas that are caused by the chemical action of the acid. If these are not removed, they will coat the carbon, insulating it from the rest of the cell. Nowadays, nearly all such cells are available in 'leakproof' form, using an outer layer of steel which is more resistant to the attack of the mild acid.

Many other cell types are used now, including the manganese alkaline, the mercury cell, the silver-oxide cell and others. All of these are **primary cells**: they obtain their action from a chemical action, and when the action is complete, the cell is thrown away. **Secondary cells** use a chemical reaction which is reversible. When the EMF of a secondary cell falls to a low value, the cell can be connected in a circuit to a supply at a higher voltage. This has the effect of reversing the chemical action, and so allowing the cell to work again. The most common type of secondary cell is the lead-acid type used in the car battery. The nickel cadmium type is more familiar in electronics work, and is also used in a variety of cordless power tools.

Whatever type of cell is used, its EMF is low, around 1.5 V for a primary cell, up to 2 V for a secondary cell. To obtain higher EMFs from cells, we need to connect them together, making a **battery**. Connecting cells positive to negative, as in Figure 3.3(a), is called *series* connection, and results in a higher EMF being

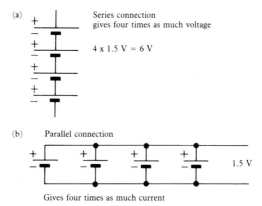

Figure 3.3 *Series (a) and parallel (b) arrangements of cells. Parallel arrangements are seldom satisfactory because if one cell is at a slightly lower EMF, current will flow to it from the others*

obtained at the two open contacts. This EMF will be the sum of all the EMFs of the individual cells. An alternative is to connect several cells positive to positive and negative to negative as in Figure 3.3(b). This is *parallel* connection, and the result is a battery whose EMF is the same as that of one cell, but which can supply much more current.

Conducting paths

A circuit is a conducting path for current, and is surrounded by insulating material so that current cannot take any alternative paths. In such a circuit, no current is ever unaccounted for. All of the current that leaves one terminal of the source (battery or generator) will return to the other terminal; current cannot be lost from a DC circuit. In addition, if the current passes through several electrical devices in turn, each device will have a potential difference across it, caused by the flow of current. The sum of all these potential differences in a current path must equal the EMF that is driving the current round the circuit. These two principles, put into a more mathematical form, are known as **Kirchhoff's Laws** (Figure 3.4), and they apply to steady currents and voltages in any DC circuit.

Effects of current

Electric current has three very noticeable effects by which we can measure it and which we can make use of. One is *heating* effect, also called Joule heating. When an electric current, steady or fluctuating, passes through any material, heat is generated. This heat is caused by the resistance of the material to the flow of current, and the amount of heat energy that is released is exactly equal to the amount of electrical energy that is lost from the circuit. The heating effect is used in all types of electrical heaters, but for many purposes it is an undesirable side-effect of electric current, and we try to use conductors that offer as little resistance to the flow of current as possible in order to reduce the losses due to heating.

The magnetic effect is even more useful. When electric current flows, there is a field generated around the conductors that affects other currents, and also some materials such as iron and steel.

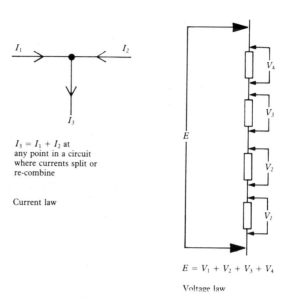

Figure 3.4 *Kirchhoff's Laws*. In any circuit, current is neither created nor destroyed, so that the total current out of a point where conductors join is equal to the total current in, and all the current from a cell will return to the cell. The potential differences across parts of a series circuit will also add up to the value of the EMF of the cell

This is a magnetic field, and its effects are the same as those that occur around a permanent magnet. The important difference is that the magnetic field around a conductor exists only for as long as the electric current is flowing, so it can be switched on and off, and varied in size. The magnetic effect is the basis of electric motors, relays and solenoids, inductors and transformers, and the conventional types of loudspeakers.

The *chemical* effect occurs mainly when electric current flows through a solution of materials in water. The effect of the current is to make ions move, and the chemical effects happen when the ions are discharged. This effect is used in electroplating and electropolishing, but it can have unwanted side-effects, such as causing corrosion of conductors in damp conditions.

Measurement of current

Current can be measured by its heating effect (still used in hot-wire ammeters), by its chemical effect (once used as a standard for current measurement), but is nearly always measured nowadays by its magnetic effect. The details of measurement methods are dealt with in Chapter 7, but the basic principle makes use of a coil and a permanent magnet. When current flows through the coil, the coil becomes a magnet which will be acted on by the permanent magnet. This force is used to rotate the coil against a spring, and a needle fastened to the coil indicates a distance over a scale. The advantage of this type of meter is that it can be manufactured comparatively cheaply, is reasonably robust, fairly precise, and easy to use. Such meter movements are the basis for practically all the non-digital multimeters that are used.

For electronics purposes, it is often important to measure very small currents. A power supply may provide a current of several amps, but individual parts of the circuit often take currents of a few milliamps (mA) or even microamps (μA). Modern measuring instruments therefore have to cope with this range of currents. When a current meter is used for DC, the terminals of the meter have to be connected in series with the circuit whose current is to be measured. This means breaking the circuit (Figure 3.5) so as to connect the meter, and this is often undesirable and sometimes impossible. For that reason, current measurements are made only where nothing else will do. The measurement of current can also be time-consuming. For most electronics purposes, measurements of potential difference (voltage) are much more useful, provided that you know how to make use of them and under what conditions they can be relied on. We will look at that idea more closely in Chapter 4.

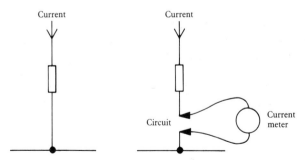

Figure 3.5 *A circuit must be broken in order to measure current, and in modern circuit boards this can be very inconvenient*

Ohm's Law and circuits

A fundamental and useful principle, called **Ohm's Law**, is the basis of much that we need for working with DC circuits. Unfortunately, the law is often misquoted, so in this section we'll look at what it really means and how it can be applied. To start with, we have to think about what is meant by electrical resistance.

Electrical resistance means the resistance of a conductor to the flow of electricity, and all conductors, apart from a few materials at very low temperatures (superconductors) have resistance. The resistance of a conductor is measured by the size of the potential difference (PD) across the conductor when a current flows. If this PD is V volts when the current is I amps, then the resistance is V/I ohms, the ohm being the unit of resistance, the volt-per-ampere. This relationship, written as $R = V/I$, $V = RI$ or $I = V/R$, is often called Ohm's Law, but it's not; it is simply the formula that defines what we mean by resistance.

Ohm's Law is more subtle and more useful. It states that for a metallic conductor at a constant temperature, the resistance is constant. This means that in the equation $V = RI$, or any of the other formulations of this equation, R is constant even though V and I can vary. For example, if the resistance of a conductor is 2 ohms, then a current of 1 A will cause a PD of 2 V, and a current of 2 A will cause a PD of 4 V. If the resistance were not constant, we could not assume that the value of 2 ohms that we used for the 1 A current could also be applied to the 2 A current. That's what Ohm's Law is about. It is particularly important to recognize this in electronics, because we make use of a lot of devices and materials which do not obey Ohm's Law. For these, we have to express voltage and current in a graph, or by measuring different values of resistance for different conditions of current and/or voltage.

For many electronic components, Ohm's Law is obeyed, and we can measure a value of resistance that we can use under a wide range of conditions. Circuits that contain only resistance of this type are called **resistive circuits**, or ohmic circuits, and the only factor that will change resistance values is temperature. The effect of temperature on a resistive material is fairly small, a change of about 0.36 per cent of value for each degree Celsius. For example, if the measured resistance of a coil of wire is 1200 ohms at 20 °C, then at 30 °C it will have changed by 0.36 × 10 × 1200/100 ohms, which is about 43 ohms. This is a small change from a starting value of 1200 ohms, but for some purposes it might be important and it might be necessary to keep components at a constant temperature. For metal conductors, and for some others, notably carbon, the change of resistance is *positive*, meaning that as the temperature increases, the resistance value increases. The **temperature coefficient of resistance** is the quantity that measures the change, and it can be quoted as a percentage change per degree Celsius, as shown above, as the actual change of resistance divided by starting resistance, per degree Celsius or in terms of parts per million of change per degree. This very common way of expressing temperature coefficients makes the typical figure 3600 parts per million (ppm) per degree. Most of the materials used in electronics for resistor construction have a very much lower figure for temperature coefficient, of the order of 100 ppm. For many purposes, resistance changes caused by temperature are ignored, and as we shall see, the changes in semiconductors and other non-ohmic conductors are very much greater.

The resistance of a conductor is due partly to the type of material, but more so to the dimensions of the conductor. A slab of metal will have low resistance if it is in the form of a short thick rod, but the same metal would have a much higher resistance if it were in the form of a long thin wire. The effect of dimensions and material on resistance is expressed by the formula shown in Figure 3.6. In this formula, the resistance is affected by length, area of cross-section, and **resistivity**. The resistivity is specific to

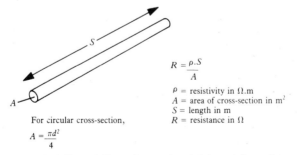

Figure 3.6 *The effect of dimensions and resistivity on the resistance of a sample of material*

the material, and Figure 3.7 shows some typical values, all in units of ohms.metres (not ohms per metre). The formula allows the calculation of the resistance of a length of wire whose material, length and area of cross-section are known, and this is sometimes of use when a precise value of (low) resistance is needed.

Material	Resistivity in $\Omega m \times 10^8$	Material	Resistivity in $\Omega m \times 10^8$
Aluminium	2.7	Steel	18
Copper	1.72	Brass	6
Lead	21	Constantan	49
Mercury	96	Manganin	45
Molybdenum	5.6	Nichrome	112
Nickel	7.8	Phosphor-	
Silver	1.6	bronze	9.4
Tungsten	5.5		

Figure 3.7 *Typical figures of resistivity of some common materials. Elements are listed on the left, and some common alloys on the right. All figures have been multiplied by 10^8, so that the true resistivity of aluminium, for example, is 2.7×10^{-8} ohm.metre*

Resistors for electronics use are mainly of thin-film construction. These employ a former such as a glass or ceramic rod on which is deposited a thin film of metal or carbon. A screw-thread pattern is then cut into the film so as to create a thin narrow strip of material between the ends of the rod (Figure 3.8), and the connections are made to these ends. The resistance value can be controlled closely using this technique, and the older methods based on a mixture of carbon and clay (the same mixture as in pencils) are used for resistors whose value tolerance can be very large, and which can have large temperature coefficient values. Metal film resistors can easily achieve tolerances of 5 per cent, and temperature coefficients of 0.01 per cent (100 parts per million) or less.

Voltage drop and dissipation

When a current flows through a resistor, there will be a potential difference (or voltage drop) across the terminals of the resistor. This applies not only to resistors but to any conductor that has

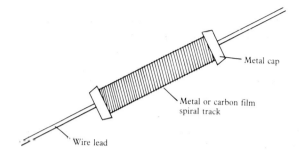

Figure 3.8 *The construction of a film resistor. The whole body of the resistor will be coated with an insulator after processing*

resistance, such as coils of relays and loudspeakers. The amount of voltage drop in a DC circuit is as given by the equation $V = RI$, the product of resistance value and current. Some care has to be taken over units when this is calculated, however. If current is in amps and resistance in ohms, then the units of voltage are simply volts. If the current is, as is more usual in electronics, in milliamps, then the result will be in volts only if the resistance value is in kilohms. Similarly, if the current is in microamps, the voltage will be in volts only if the resistance value is expressed in megohms. Figure 3.9 is a reminder about the units for this equation.

Units of V	Units of R	Units of I
V	Ω	A
mV	Ω	mA
kV	k	A
V	k	mA
mV	k	μA
kV	M	mA
V	M	μA

μ (micro) = $\frac{1}{1\,000\,000}$ or 10^{-6}

m (milli) = $\frac{1}{1000}$ or 10^{-3}

k (kilo) = 1000 or 10^3
M (mega) = 1 000 000 or 10^6

The Ω sign is often omitted when k or M is used, and in values such as 2.7Ω. The letter R may be used in place of the decimal point so as to avoid the Greek letter Ω, hence 2R7, 3R3 etc.

Figure 3.9 *The units of voltage, resistance and current that will be encountered in electronics*

The flow of current through a resistor also causes power to be lost in the form of heat dissipation. Power is the rate of working, the amount of work done per second or, as in this case, the amount of energy lost per second. If the voltage drop across a resistor is known and the current through the resistor is known also, then the power loss is given in watts by $V \times I$. Figure 3.10 shows the alternative forms of this equation, which can be used when only resistance and current or resistance and voltage are known. These equations, like most of the equations of circuits, apply only to the parts of the circuit that obey Ohm's Law.

$$\text{Power} = V \times I$$
$$= \frac{V^2}{R}$$
$$= I^2 R$$

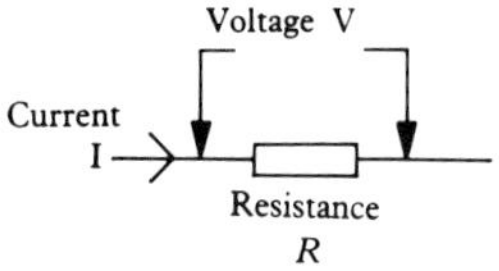

Figure 3.10 *The three versions of the power formula. For electronics use, it is more usual to know V and R so that the V^2/R version is more useful.*

The allowable power dissipation is a very important consideration for a resistor, because it determines the temperature rise of the resistor when it is used. There is no simple way of calculating the temperature at which a resistor will run at any particular level of power dissipated in it. This is because the temperature that a resistor reaches depends on how fast the heat can be carried away by the air. The temperature of the resistor will become steady when the rate of removing heat is the same as the power dissipated, and this depends on how tightly the resistor is packed in the circuit, the presence of other resistors, the air flow over the resistor and so on. Most resistors have their ratings determined at a temperature of 70 °C, which is the 'average' temperature of operation in a circuit with the resistor at full rated dissipation. The temperature can rise considerably higher than this if the layout of a circuit is cramped, so that if a resistor runs very hot and fails as a result, it's wise to replace it with a resistor rated for higher dissipation, or to arrange for better cooling. A surprising number of electronics failures arise simply from overheating of resistors and capacitors.

4
DC circuits

The electrical circuit consists of a conducting path between the terminals of a supply, and various devices may be connected in this path. In electronic **circuit diagrams** (known in the US as *schematics*), the connections of the circuit are represented by solid lines, and the devices by symbols, some of which are illustrated in Figure 4.1. The importance of the circuit diagram is that it shows very clearly how the circuit will work, independent of the way that the circuit might be laid out in practice. The practical circuit layout is shown in a **layout diagram**, and this type of diagram is useful for fault-finding, because it makes the location of each component easier. The circuit diagram, however, is the one that is needed in order to construct circuits and to carry out testing. From this point onwards, we'll use the standard symbols for circuit diagrams that explain the working of a circuit or of components. For a DC circuit, the only calculations that are likely to be needed are these based on $V = RI$.

Series and parallel

The two fundamental connections of components in a circuit are series and parallel. In a series circuit of resistors, Figure 4.2, the same amount of current flows in each resistor. A set of resistors in series offers the same total resistance to a current as a single resistor whose value is equal to the sum of the series resistors. The alternative arrangement of parallel resistors is shown in Figure 4.3. In this case, the voltage across each resistor is the same, but the current through each is given by $I = V/R$ as usual. A set of

38 Electronics for Electricians and Engineers

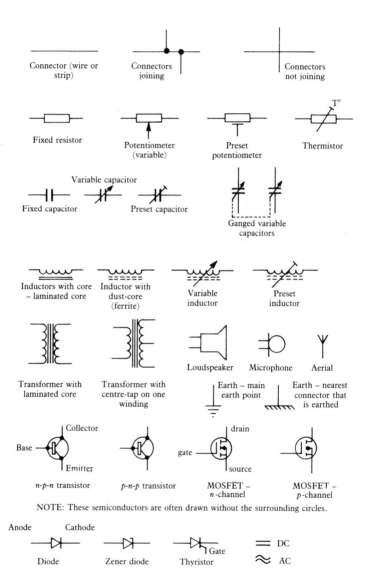

Figure 4.1 *A selection of the most common circuit symbols in international use*

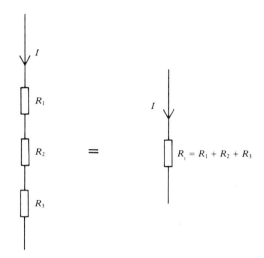

Figure 4.2 *Resistors connected in series. The total value is the sum of the individual values provided that the same current flows in each resistor.*

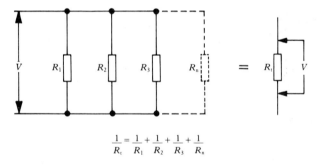

Figure 4.3 *Resistors connected in parallel. The equation applies only if the same voltage exists across each resistor.*

resistors in parallel takes the same current as, and is therefore equivalent to, a single resistor R_t whose value is given by:

$$\frac{1}{R_t} = \frac{1}{R_1} + \frac{1}{R_2} + \frac{1}{R_3} \ldots$$

For the more usual arrangement of two resistors, R_1 and R_2 in parallel, this can be simplified as shown in Figure 4.4.

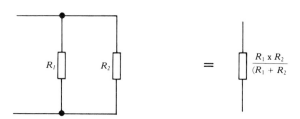

Figure 4.4 *The common case of two resistors in parallel and the resultant value.*

Preferred values

You might imagine that resistors would be manufactured in all possible values, but this would create chaos in terms of storekeeping. The compromise is to manufacture for certain preferred values, using figures that are fixed for different tolerance sets as listed in Figure 4.5. In each of these sets, the numbers lie between 1 and 10, and the letters R, K, and M are used to indicate units, thousands and millions respectively. In modern usage, the letters are put in place of the decimal point, so that 4K7 means 4700 ohms, 3R3 means 3.3 ohms, and 1M5 means 1 500 000 ohms. In each of these sets also, the values overlap slightly, so that any actual value of resistance *must* fit into a value on the set and be within the tolerance. For example, in the 20 per cent tolerance set, we have the figures 3.3 and 4.7. 20 per cent of 3.3 is 0.66, and so a resistor whose nominal value was 3.3 K could have a value as high as 3.3 + 0.66 K = 3.9 K. Similarly, a resistor marked as 4.7 K with 20 per cent tolerance could be of a value as low as 4.7 − 0.94 = 3.76 K, since 20 per cent of 4.7 is 0.94. This means that a resistor that turned out to have a value of 3.9 K could be sold either as a 3K3 or a 4K7. In practice, of course, it would be sold as a 3K9 10 per cent tolerance resistor!

Resistors, like other components, are manufactured so as to aim at the preferred values, and then sorted by machine into ranges. The values which fall very close to the preferred values can be sold as 1 per cent or 5 per cent tolerances, some others as 10 per cent, and the rest as 20 per cent. In other words, resistors have been sorted before you get them, and there is no point in trying to search through a box of 4K7 resistors looking for one that is 3K9, because any of that value will have been removed.

20%	1.0			1.5			2.2			3.3			4.7			6.8	
10%	1.0	1.2		1.5	1.8		2.2	2.7		3.3	3.9		4.7	5.6		6.8	8.2
5%	1.0	1.1	1.2	1.3	1.5	1.6	1.8	2.0	2.2	2.4	2.7	3.0	3.3	3.6	3.9	4.3	4.7	5.1	5.6	6.2	6.8	7.5	8.2	9.1

Figure 4.5 The preferred set of values for 20 per cent, 10 per cent and 5 per cent tolerance. These values are also used for capacitors and for the voltages of zener diodes

Potential dividers

The **potential divider** is a circuit that is used to a very large extent in electronics, perhaps more than any other form of DC circuit. The simplest form of potential divider consists of two resistors connected in series between the positive and the negative terminals of a power supply (Figure 4.6). The purpose of the circuit is to obtain a voltage level V that is lower than the power supply voltage E, by making use of the voltage drop across the resistors. The analysis of the circuit shows that the voltage across the resistor R_2 is given by the expression:

$$\frac{E.R_2}{R_1 + R_2}$$

so that the voltage out is a fixed fraction of the supply voltage. This assumes that the current taken from a point X to whatever is connected there is negligible. If the current taken from X, by whatever device is connected there, is not negligible compared to the current flowing through R_1 and R_2, then the voltage V will be lower than the formula predicts. The applications of the potential divider in electronics practically always ensure that this assumption will be justified.

The use of the potential divider avoids having to supply a large variety of different voltages to a circuit, so that a supply of, say, +5 V or +12 V can provide for all the voltages of +1 V, +1.6 V, +2.2 V and so on that might be needed in a circuit, as long as these require negligible current. A good definition of negligible current is a current less than 5 per cent of the current through the resistors in the potential divider. In some cases, this may mean that the resistors R_1 and R_2 may have to handle several milliamps of current, in other cases, less than 1 mA will be adequate. If the

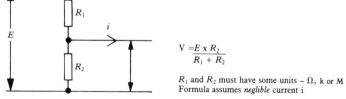

Figure 4.6 *The potential divider and the formula for the divided voltage* V

resistors have to be able to pass a substantial current, the power ratings may have to be more than the usual 0.125 W or 0.25 W rating of small metal-film resistors.

Another application of the potential divider circuit is to variation of voltage. We might, for example, be working with a 12 V supply, and need a voltage that could be varied between 0 and 12, or perhaps between 1 and 2 V. Such a supply can be obtained by using a **potentiometer**, either as a variable or as a preset. The potentiometer is a resistor with three terminals. One fixed terminal is connected to each end of the resistive material, and the third terminal is moveable, and can make contact with any part of the resistive material. The conventional form of potentiometer has the resistive material, such as a metal film, deposited on a circular ring over about 270 degrees, with a connection to each end. The variable contact, or tap, consists of a piece of springy metal rubbing against the film, and connected to a third terminal by a stranded wire 'pigtail' or by another sliding contact bearing on a metal ring to which the third terminal is connected (see Figure 4.7). The symbols are shown in Figure 4.8, with the arrowhead indicating a control that will frequently be used (like a volume control) and the 'T' shape indicating a preset which is fixed at the time of setting up the circuit and thereafter adjusted only if necessary.

The connection of a potentiometer across the power supply lines, as shown in Figure 4.8, allows the voltage at the output terminal to be adjusted from zero up to the full voltage of the supply. The potentiometer must be of a value that its power rating can cope with. If, for example, the supply voltage is 12 V, then a 120 R potentiometer will dissipate a power of 12 × 12/120 = 1.2 W. That would rule out the use of 0.25 W, 0.5 W or even 1 W potentiometers. A value as low as 120 R would be used only if the current taken from the tap terminal needed to be around a

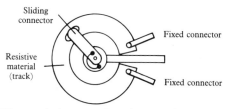

Figure 4.7 *The practical construction of a potentiometer*

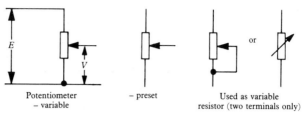

Potentiometer — variable   — preset   Used as variable resistor (two terminals only)

Figure 4.8 *Symbols for potentiometers, variable or preset*

milliamp, and we would more usually employ a 10 K value if the current were negligible. Presets are generally of low power ratings, and use ceramic and metal film construction (cermet presets). The traditional carbon track potentiometer is still used, and higher power ratings can be obtained by using wire-wound potentiometers.

In many circuits, a variation from zero to full supply voltage is very undesirable, and the range of the potentiometer must be limited. This is done by connecting resistors in series, as Figure 4.9 shows. The values of the potentiometer and the resistors will be chosen so that the voltages at the fixed terminals of the potentiometer are the appropriate voltages for the range that is required. The example shows values that have been calculated for a 12 V supply required to give a range from about 1 to 2 V. As always, using resistors of the nearest preferred values means that the voltages will not be exact, but this is of no great importance when the output from the circuit is adjusted by means of the potentiometer in any case.

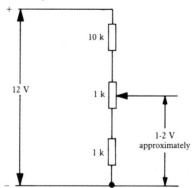

Figure 4.9 *Connecting resistors in series with either end of a potentiometer in order to limit the range of voltage*

The bridge circuit

Another common DC circuit, also used for AC, is the **bridge**. The bridge is usually drawn as in Figure 4.10a, but the diagram of Figure 4.10b is clearer. This shows that the circuit consists of two potential dividers, and the voltage marked V is the voltage between the two tapping points of the dividers. The theory of the circuit is shown in the diagram, but the conclusion is the important point. In nearly all applications of this type of circuit, the aim is to adjust the division ratio of one divider so that the voltage V is zero, a condition referred to as balance. This is achieved when the values of the resistors are given by:

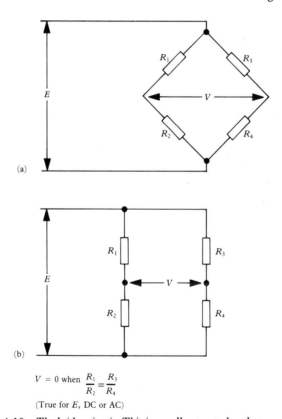

$V = 0$ when $\dfrac{R_1}{R_2} = \dfrac{R_3}{R_4}$

(True for E, DC or AC)

Figure 4.10 *The bridge circuit. This is usually operated so that one potential divider is adjustable. The aim is to make voltage V equal to zero, when the bridge is said to be balanced*

$R_1/R_2 = R_3/R_4$, and this can be simply achieved by making one pair fixed resistors, and the other pair a potentiometer whose ratio can be varied.

The division ratio of a potentiometer can be marked on its scale, so that the circuit provides an excellent method of measuring resistance. Suppose, for example, that R_2 is a known 1 K resistor, and R_1 is unknown. If the resistors R_3 and R_4 are represented by a potentiometer, then this can be adjusted until the circuit is balanced, meaning that the voltage V is zero. Detecting a zero voltage is very simple, because the meter that is used need not be calibrated. If we find that the potentiometer ratio is 2.66 in this balanced condition, then the value of the unknown resistor is 2.66 times the known 1 K resistor, that is 2K66. This bridge circuit is the basis of many popular measuring instruments for resistors, capacitors and inductors, and is used with AC as well as with DC.

Measuring instruments

Many voltage measurements on circuits are of the voltage produced by a potential divider, and this is a good example of how the use of a measuring instrument can alter a voltage. Connecting a voltmeter to a potential divider (Figure 4.11) is equivalent to connecting the resistance of the voltmeter across one of the resistors in the divider chain, so making the resistance of this part of the chain lower than it was. As always, if the

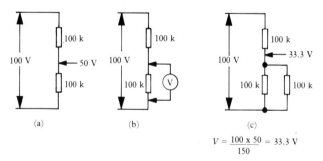

$$V = \frac{100 \times 50}{150} = 33.3 \text{ V}$$

Figure 4.11 *The effect of using a voltmeter. The potential divider (a) has 50 V at its centre point. Using a voltmeter (b) to measure this will connect the resistance of the voltmeter in parallel with one resistor (c). This makes the voltage lower than the voltage that exists when the voltmeter is not connected*

resistance of the meter is much higher (at least ten times higher) than the resistor in the chain, the effect will be small, but if the voltmeter resistance is low, the reading will be false. As indicated in the previous chapter, the resistance of a meter or any other instrument should be known so that you can estimate the effect that it will have on a potential divider or any other circuit.

Equivalent circuits

A DC circuit may consist of a large number of components connected together with several current paths. We have seen that a set of resistors in series can be represented by one resistor, and a set of resistors in parallel can also be represented by one resistor of suitable value. In the same way, any number of components in a DC circuit can be represented by a much simpler circuit, called the **equivalent circuit**. This is true also of AC circuits, but the equivalent circuit for AC is not usually identical to the equivalent circuit for DC.

The simplest equivalent circuit consists of a supply, a series (or source) resistor, and load resistor. This equivalent circuit can be used for any DC circuit, though the values of the EMF, and of the source and load resistors will be different in each case. The benefit of the equivalent circuit, if its values are known, is that it makes it very easy to calculate the effect of changes. Suppose, for example, that the equivalent circuit is as in Figure 4.12. You

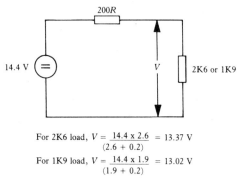

For 2K6 load, $V = \dfrac{14.4 \times 2.6}{(2.6 + 0.2)} = 13.37$ V

For 1K9 load, $V = \dfrac{14.4 \times 1.9}{(1.9 + 0.2)} = 13.02$ V

Figure 4.12 *Using an equivalent circuit. Any circuit that contains a source of voltage and resistors can be simplified to a voltage supply, a series resistor and a load resistor. This greatly simplifies calculations on the voltage and current at the load*

cannot expect the resistors of this theoretical equivalent to have normal preferred values, so that the figures are bound to look odd. From this equivalent, the voltage across the load can be calculated by the normal potential divider equation as about 13.37 V. Now if a change to the real circuit makes the equivalent change so that the load resistance value becomes 1K9, then the voltage across it will become 13.02 V. The only circuit theory that you need here is that of the potential divider.

The equivalent circuit we have used is that of a voltage generator and internal resistance. The voltage generator is the source of voltage V, and the internal resistance is the series or source resistance. Neither of these necessarily corresponds to any real voltages or resistance, but in most cases, V might be the EMF of a battery of power supply, and the source resistance would be the internal resistance of the battery or power supply. Both of these factors would be fixed by the design of the battery or power supply, though as a battery ages its internal resistance rises. The electronic circuit itself would then be totally represented by the load resistance. Equivalent circuits are often used for parts of a circuit, however, in which case the values do not correspond in such an obvious way to values that can be measured. This is very much more important when we come to consider AC equivalent circuits.

Thevenin's Theorem

The name of equivalent circuits is made much easier by an important theorem that bears the name of its discoverer, Thevenin. Thevenin's Theorem states that any linear circuit (i.e. one in which all components obey Ohm's Law) can be represented by the simple equivalent circuit of a supply voltage, source resistance and load resistance. The theorem then goes on to show how these quantities can be calculated. This is not something that you need to use very often, and it can be difficult for very complicated circuits, but some knowledge of how the theorem is applied can be useful. The principle is to imagine the circuit broken at some point where you want to take readings. The load resistance is the resistance on the 'cold' side of the break, the side that is no longer connected to the power supply. The equivalent voltage is the voltage at the other end of the break, and the series resistance is the resistance from the break to earth, *imagining that the power supply is a short circuit.*

An example is useful at this point. Suppose we have the circuit of Figure 4.13, and we want to know how the voltage at point X will change as we alter the value of R_5. The equivalent circuit will have to be constructed by imagining the circuit broken at point X, as illustrated in Figure 4.14a. We can see that R_5 is the load resistance of 200 ohms, but we need to calculate the values of V and the source resistance. Finding the value of V is easy when the remaining circuit is simplified by replacing R_1 and R_2 (Figure 4.14b) with a single resistance having the value of R_1 and R_2 in series. The circuit now becomes a potential divider giving 6 V at its output, so that V in the equivalent is 6 V.

The more tricky calculation is that of series resistance. If we imagine the power supply shorted, the circuit looks like that of Figure 4.14c, consisting of 100 ohms in series with a parallel combination of two 300 ohm resistors. As the diagrams show, this is equivalent to a resistor of 250 ohms, using only the simple series and parallel resistor replacements. The equivalent circuit then becomes as in Figure 4.15, consisting of a 6 V supply, a series resistance of 250 R, and the 200 R load. As it stands, the voltage at point X will be 2.67 V, and the voltage for any other value of R_5 can be calculated by making use of the simple straightforward potential divider equation. Contrast this with the calculation that you would need if you were to analyse the circuit completely each time you needed to find the voltage at X for a given value of R_5.

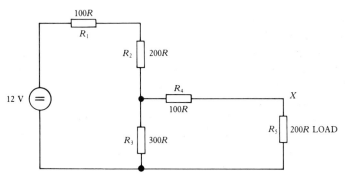

Figure 4.13 *A circuit used as an illustration of Thevenin's Theorem to simplify a circuit to that of Figure 4.12*

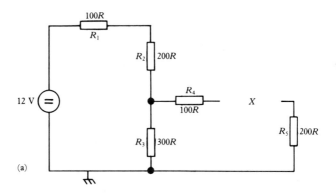

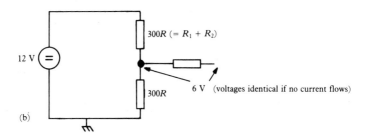

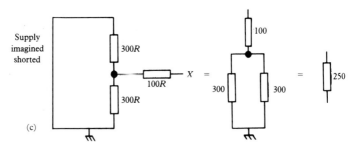

Figure 4.14 *Steps in using Thevenin's Theorem: (a) imagine the circuit cut at point X; (b) determine the voltage at the break – there is no drop of voltage across R_4 if no current flows; (c) determine the resistance from the break point back, with the power supply imagined as a short circuit*

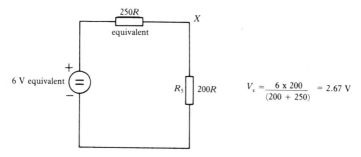

Figure 4.15 *The equivalent circuit of Figure 4.13 obtained by using the steps of Figure 4.14*

---

1 Imagine the circuit broken at the 'live' end of the load resistor.
2 Redraw the circuit so as to find the voltage at the break. Make this the voltage for the generator in the equivalent circuit.
3 Find the resistance from the break to earth, imagining that the power supply is a short circuit. Make this value the series resistor in the equivalent circuit.
4 Draw the new equivalent circuit, with its generator voltage and series resistor. Connect in the load, and calculate the load voltage.

Figure 4.16 *The rules for using Thevenin's Theorem summarised. These can be applied to AC circuits also, but not so simply*

Superposition

Thevenin's Theorem is very useful for DC circuits with just one voltage supply. Occasionally we need to consider what happens in a circuit that contains more than one supply. This may be because the circuit contains two separate power supplies at different voltages, or because a back-up battery is used, or because a large capacitor is charged so as to act as a power supply for part of the circuit. Whatever the reason, the normal 'textbook' method of dealing with such circuits is by use of Kirchhoff's Laws. Unfortunately, most schools stopped teaching, many years ago, the algebra that is needed for solving Kirchhoff's Laws (and most other engineering calculations). The options you are left with are of using a maths textbook intended for some other country, or of using the **Superposition Theorem** for the problem.

52 Electronics for Electricians and Engineers

The Superposition Theorem looks relatively easy if you have previously used Thevenin's Theorem. Circuits that contain two supplies can always be analysed to an equivalent such as is shown in Figure 4.17: this example shows values that will be used to demonstrate the Superposition Theorem in action. The problem is to find what the voltage across *XY* will be due to the two supplies. The Superposition Theorem asks you to imagine that this voltage is the sum of two voltages, one due to each supply. The voltage due to a supply is simply the voltage that would exist if the other supply were a short circuit. Figure 4.18 shows these two possibilities, giving voltages of 2.4 V and 4.0 V respectively. The sum of these is 6.4 V, which is therefore the voltage across *XY*. The advantage of using the Superposition Theorem is that all the arithmetic is simple, depending only on forming the result of series and parallel resistors.

Even if no circuit diagram is available, it is possible to obtain the equivalent resistance of a circuit by practical methods. Once again, we are assuming that we are dealing with a DC circuit, but the methods are pretty much the same when AC circuits are being used. The scheme is to measure the voltage across the part of the circuit which is regarded as the load, along with the current through the load, and to measure how the voltage changes as the current is changed. At least two readings of voltage and current will be needed, and Figure 4.19 shows how the values of voltage and resistance are found from the readings.

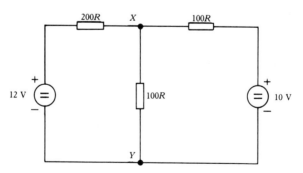

Figure 4.17 *A type of two-supply problem that is better tackled with the Superposition Theorem*

DC circuits 53

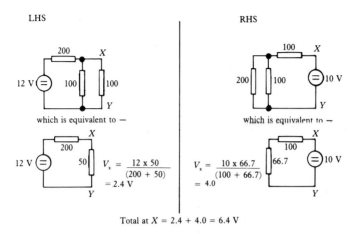

Figure 4.18 *The analysis of the circuit of Figure 4.17, using the Superposition Theorem. The circuit is treated one supply at a time, with the other supply imagined shorted. The method can also be applied to AC circuits with some modifications*

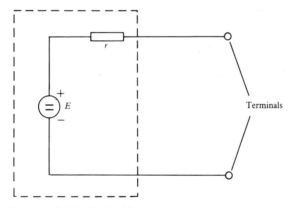

'Components' within the dotted lines cannot be measured directly

1. Using a high resistance voltmeter, measure the voltage at the terminals. This gives the value of E.
2. Keeping the voltmeter connected, connect a load so that the voltage reading drops by an easily measurable amount. Record this voltage reading, V and the load resistance R.
3. Find r from the expression: $r = \dfrac{R(E - V)}{V}$

Figure 4.19 *How to calculate E and r for an equivalent circuit from direct readings at two terminals of a circuit*

Non-linear resistance

So far, we have assumed that a circuit will contain only components that obey Ohm's Law, i.e. linear components. The most important of the electronics components do not obey Ohm's Law, and quite a number of other components, notably thermistors (Figure 4.20), are also non-linear. How then do we deal with such components in a circuit? It is not such a problem as it might appear because no circuit is constructed from components that are all of this type. The usual type of electronic circuit contains one component that is non-linear, but the currents and voltages are controlled by familiar linear components such as resistors. Because of this, we very seldom need to know in detail how a non-linear component behaves, because we can work out the voltages and currents from the linear components.

As an example, a transistor of the old-fashioned bipolar type will require a voltage to be applied between two terminals, the

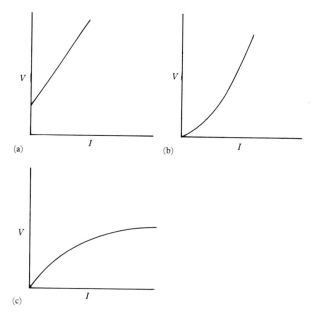

Figure 4.20 *Non-ohmic behaviour (a) of an electrolytic cell in which no current can flow until a reverse voltage is overcome, (b) of a thermistor with positive temperature coefficient, (c) of a thermistor with negative temperature coefficient*

base and the emitter. These terminals form a non-linear connection, with a relationship between voltage and current that is nothing like that of a resistor. We can, however, fix the voltage at the base terminal by using a potential divider, and the current through the emitter terminal by using a series resistor. In this way, any calculations that have to be made are of the voltage output from a potential divider and the current through a resistor, both familiar $V = RI$ calculations. The non-linear behaviour of the transistor terminals does not matter. When we measure voltages, we are simply measuring the voltage across resistors in straightforward resistive circuits, and once again, the action of the transistor does not affect the way that we take the readings.

Finally, there is an alternative theorem to Thevenin's, called Norton's Theorem. This is very similar, but makes use of a current supply and parallel resistors rather than a voltage supply and series resistance. It is more useful to the circuit designer than to the service engineer or technician, and has been omitted here.

5
Electricity and heat

Ever since heat was found to be a form of energy, the conversion of other forms of energy into heat has been of interest, as well as the much more difficult problem of converting heat into other forms of energy. Without going into detail, it's always easy to convert any other form of energy into heat; in many cases it's difficult to avoid. Converting in the other direction is much more difficult, and always involves losses of energy because heat can flow only from a high temperature to a low temperature. Any conversion of heat into other forms of energy therefore involves taking in heat at a high temperature and giving some exhaust heat out at a lower temperature. This exhaust heat is lost, and cannot be used in the conversion.

Heating effect

Electrical energy is converted into heat energy whenever a current flows through a conductor that has resistance. Since all materials, other than superconductors, have electrical resistance, this means that electrical energy is being converted into heat whenever current flows. The conversion to heat causes the temperature of the conductors to rise, so that heat then flows in the conductor towards anything at a lower temperature, usually the air. This flow of heat represents a loss of energy, so that no electrical circuit can make 100 per cent use of the energy that is supplied. There is even a loss within a generator (battery or alternator) itself caused by the resistance of the conductors such as electrolyte or wire windings.

Power is defined as the rate at which energy is converted, meaning the amount of energy converted per second. As it happens, the units that are used for electrical current and potential difference are closely linked to power, as we have seen. For a conductor with a current I flowing across a potential difference of V volts, then the power is given by VI. The unit of power is the **watt**, and this in turn represents a change of energy of one joule per second. The **joule** is the unit of energy or work, no matter what form it takes. For electrical work, the energy in joules is the power in watts multiplied by the time in seconds for which the power is converted. Since the conversion of electrical power or energy into heat nearly always results in the heat being dissipated into the air, we speak of this as power dissipation or energy dissipation. For most electrical purposes, other than calculating the cost of electricity, power is a much more important and useful quantity than amount of energy, so that the watt is a more common unit than the joule. The conversion of electrical power into heat is made use of in some devices, and in others is a drawback that has to be minimised. For the rest of this chapter, we shall be looking at the applications of this conversion, and how its disadvantages can be minimised.

Heating elements

Heating elements are conductors that have a comparatively high resistance, so that electrical power will be deliberately converted into heat. The most common material for this purpose is nichrome, an alloy of iron, nickel and chromium. This material has a high resistivity, making it ideal for all purposes where high-resistance wire windings are needed. It has also a high melting point, and is resistant to oxidation at high temperatures, so that it can be used for heaters operating up to red-hot temperatures. Virtually all common heating elements make use of nichrome, and nichrome wire is also used for the manufacture of wire-wound fixed resistors and potentiometers. When a heating element is part of an electrical circuit, some care has to be taken about any circuit calculations that are made if the temperature range is large. In electronics, heating elements are often intended to maintain only a small temperature difference above normal air temperature, and no special allowances have to be made, but for a heater operating at a high temperature, several hundred degrees C, then any calculations must take account of the resistance that the material will have at its normal operating temperature, which will certainly not be the same as the resistance at room temperature. Fortunately, if nichrome is used, the resistance change is smaller than it would be for some other materials.

For example, suppose that a heater is to operate at 240 V, 2 A. This implies that its resistance will be 120 ohms, but this is the hot resistance value. When you use an ohmmeter to measure the resistance of the element at normal temperature, the value that you will find will be lower than 120 ohms. How much lower depends on the temperatures involved. If the heater normally runs at 400 °C, and we take the resistance measurement at 25 °C, then the difference can be calculated from the temperature coefficient of resistance of nichrome, which has the value of 0.017 per cent. Because nichrome has an unusually small temperature coefficient of resistance, the change will be small, to about 113 ohms in this example. A heating element made of pure nickel would have a resistance of about 47 ohms, using the same resistance and temperature example.

For temperature changes of a few hundred degrees then, using nichrome, the change of resistance is not very significant, and this is one reason for using this and other nickel-iron alloys in heating elements. The same reasons apply to the use of nichrome for potentiometers and wire-wound resistors, because changes of resistance with increasing temperature are even more undesirable in this case. You should, however, expect that any resistance measurements made on heating elements at room temperatures are probably going to be low compared with the resistance at working temperature. Figure 5.1 illustrates the calculations that lead to the results quoted above.

Thermal switches and sensors

Thermal switches are used in electronics for two main purposes. One is as a safety device, switching a circuit off in the event of a large rise in temperature; the other is as a time delay, so that one voltage is switched on at a significant time later than another. The form of the switch is similar for both types of operation. The basic action is usually that of the **bimetallic strip** (Figure 5.2). This

Resistance-temperature formula is:

$$R_\theta = R_0(1 + \alpha\theta)$$

where R_θ is resistance at temperature θ °C
R_0 is resistance at temperature 0 °C
α is temperature coefficient of resistance

Equations are

1. $R_{400} = R_0(1 + 7.7 \times 10^{-4} \times 400)$
2. $R_{25} = R_0(1 + 7.7 \times 10^{-4} \times 25)$

Multiply out brackets, divide (1) by (2)

$$\frac{R_{400}}{R_{25}} = \frac{1.068}{1.00425} = 1.06$$

If $R_{400} = 120\Omega$, then $R_{25} = \frac{120}{1.06} = 113.2\Omega$

Figure 5.1 *The calculations for determining the amount of resistance change as a result of a change in temperature*

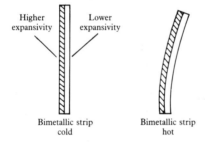

Figure 5.2 *The principle of the bimetallic strip. Despite the name, the shape need not be a strip and a circular element is usually preferred*

consists of two thin metal strips, made from metals that have widely different expansivities, meaning that their lengths will change by very different (small) amounts when the temperature is changed. By welding, soldering, or riveting these strips together, we can make a compound strip which will bend when heated. The bimetallic strip can then be used as a switch arm, carrying a contact at one end which will either make or break when the strip is heated. The usual scheme is to have the strip break the contact at some preset temperature.

For a safety switch, such as the RS Components type 339-314, the circuit is interrupted at some preset temperature. In this example, the switch contacts will open at 100 °C, with a tolerance of 4 °C either way. The contacts are rated at 250 V AC, 10 A, or at 30 V DC, 5 A, and the shape of the switch (Figure 5.3) allows it to be bolted to any flat metal surface in order to detect temperature rise. Switches of this type are used, for example, on equipment in which excessive temperature could destroy essential components, and the fact that a temperature drop of around 15 °C is needed to permit the contacts to close again provides an ample safety factor.

The other application of the bimetallic strip thermal switch is in delay switches, usually sold ready-made as time-delay relays. The principle here is that the application of power to a circuit energises a heater which is wrapped around a bimetallic strip. This strip is, as usual, part of a switching circuit, but in this case the contacts are normally open. The time delay is the result of the time needed to heat the strip to a temperature that will cause the contacts to close, and when this happens, another circuit will be energised. In some examples, this operates a holding action in which a relay takes over the circuit connection, breaking the heater supply so that the contacts of the thermal switch will open again, ready to provide another time delay if needed. Figure 5.4 illustrates the action, which is not encountered so frequently nowadays as it once was.

Figure 5.3 RS Components Ltd. *A typical temperature sensitive switch using a bimetallic element*

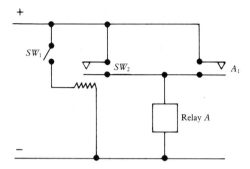

Action: The switch SW_1 is closed manually and held closed while current flows in the heater of the thermal switch SW_2. When the contacts of SW_2 close, relay A is energised, closing its contacts A_1. This holds the relay on, so that SW_1 can be released, allowing the element of the thermal switch to cool. The relay will be released when power is disconnected, and the other relay contacts can be used to switch other circuits.
For the normal nomenclature, see Figure 6.6

Figure 5.4 *Using a thermal switch with its own heater element to provide a switch-on time delay*

Thermistors and thermocouples

The **thermistor** is a form of resistor made from semiconducting metal oxides. Its temperature coefficient of resistance is very large, and by suitable choice of materials can be made either positive or negative. Thermistors are used mainly for sensing temperature changes or for measuring temperature, and the simplest method is to make the thermistor one part of a bridge (Figure 5.5). If the bridge is in balance at one temperature, any change in the temperature will alter the resistance of the

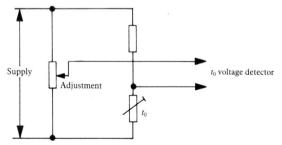

Figure 5.5 *A thermistor bridge used for temperature measurement. The reading instrument will be a voltage meter, and the potentiometer is used to provide a zero-set or for calibration to some other temperature*

thermistor, so unbalancing the bridge. The type of thermistor used for this purpose will have a negative temperature coefficient, but some types of thermistors with positive temperature coefficients are used for stabilising current. The current to a circuit is passed through a thermistor, which heats because of the current. Any increase in the current will raise the temperature of the thermistor and so increase its resistance. This in turn will make the current decrease again, so stabilising the value.

An older form of temperature-measuring device is the **thermocouple**. A thermocouple consists of two different metal wires twisted, soldered or welded together. Two such 'junctions' can be joined in series (Figure 5.6), and when this is done, there will be an EMF generated that depends on the *difference* in temperature between the two junctions. An alternative is to use one junction and connect it with a source of low voltage to simulate the other junction. The output voltage of a thermocouple is very small, but it can easily be amplified, and suitable choices of materials can allow very large temperature ranges to be measured.

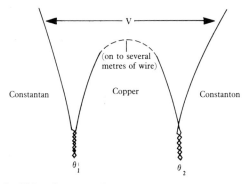

Figure 5.6 *Using thermocouples. One useful combination is copper and the alloy Constantan. The wires are twisted, soldered or welded together and when the two junctions are at different temperatures θ_1 and θ_2 there will be a voltage V between the ends. The voltage is small, typically 4 mV for 100 °C difference.*

Fuses

A **fuse** is an application of the heating effect to circuit protection. It consists of a thin conducting wire or strip made from a metal with a comparatively low melting point and high resistivity.

When a current passes through the fuse, the metal is heated, and if the current is excessive, the metal will be heated to melting point and will therefore break the circuit. Fuses are made with glass or ceramic casings so that the heat is not dissipated, and when a fuse acts, the metal is often vaporised rather than simply melted. A common construction for modern fuses is of a ceramic tube casing, filled with sand to act as a thermal insulator and also to absorb the molten or vaporised metal if a fuse blows.

The time taken for a fuse to blow with a specified excessive current will depend on the type of construction. For modern semiconductor electronic circuits, very fast-blowing fuses are needed. The speed of action is quoted as a range of milliseconds for a current of some amount more than rated current. For example, the RS Components super-quick-acting fuses will blow in a time ranging from 10 ms to 2 s for a current of 2 × rated current, and in a time range of 2–15 ms for 4 × rated current. Very fast action is not always an advantage, however. In some circuits, fuses must be able to cope with current surges of well above rated current for short times. A typical example is a surge of 10 × rated current for 20 ms, and for such applications, anti-surge fuses must be used. Such surges occur in circuits that contain inductors or electric motors, and the anti-surge fuses must not be replaced with ordinary fuses in such circuits.

Electricity distribution

The conversion of electrical power into heat, and the subsequent dissipation of heat, affects the way that we distribute electricity. Electricity must be transmitted along cables, and such cables have resistance. For a cable of resistance R, the power lost as heat will be I^2R, resistance multiplied by the square of current. Since there is no practical way of reducing the resistance of cables without making the cost of the cables uneconomic, the obvious remedy is to reduce the amount of current that flows. The use of AC for power distribution makes this possible, because AC allows voltage and current conversion by means of transformers, with very little loss. For a given amount of power, then, much more efficient transmission can be achieved if the voltage is very high and the current low. At one time, a voltage of 132 kV was standardised, but voltages of 500 kV are now in use, At 500 kV, a current of only 2 A represents one megawatt of power, and the loss of power in the cable is negligible in comparison to the amount transmitted.

Dissipation of heat

Many electronic devices need to be able to dissipate the heat that is converted from electric power. Ultimately, this heat has to be dissipated into the air, since few electronic components make use of water cooling, other than some transmitting valves. Dissipating heat into the air is much less easy than water-cooling, and it requires the heat to be passed to a **heat-sink**, a large metal surface that is in contact with cool air. The most efficient arrangement for such a surface is in the form of cooling fins (Figure 5.7), and if air can be forced over the fins the rate of power dissipation can be greatly increased.

The effectiveness of any arrangement for dissipating heat into the air is very difficult to calculate, and we generally rely on measurements made on typical examples. For any calculation of dissipation, the thermal resistance of the finned surface should be known or measured. The thermal resistance is measured in units of °C/W, meaning the temperature rise in °C above the air temperature per watt of dissipation. If, for example, a metal heat-sink is quoted as having a thermal resistance of 0.5 °C/W, this means that each watt of power converted to heat will raise the temperature of the metal 0.5 °C above the air temperature. For such a heat-sink, a dissipation of 10 W would cause a temperature increase of $10 \times 0.5 = 5$ °C above air temperature. If the air temperature happened to be 30 °C, then the dissipation of 10 W would raise the heat-sink temperature to 35 °C. Heat-sinks can be obtained in a variety of shapes, sizes and thermal resistance values, but if a heat-sink has to be made for a specific purpose, design will generally have to be by cut-and-dry methods. The thermal resistance can be found by measuring the steady temperature above air temperature attained when a resistor of specified dissipation is clamped to the heat-sink.

Electricity and heat 65

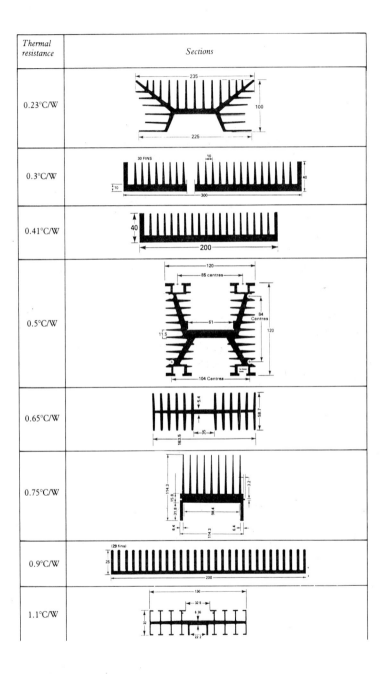

66 Electronics for Electricians and Engineers

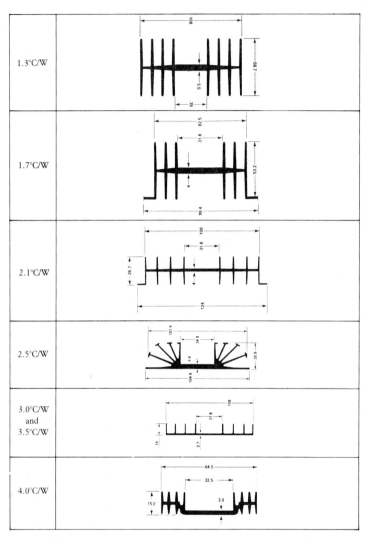

Figure 5.7 *Cooling fin patterns and their thermal resistance figures – a selection from the RS Components list*

6
Magnetism

When current flows through any conductor, and when electric field strength in an insulator changes, there will be a magnetic field. The word 'field' here has its usual meaning of some region of space where forces can act with no visible mechanical connection. The magnetic field exists around a conductor when a current flows, and its effect will be a force exerted on other conductors carrying current, and on objects made of iron (and some other metals) whether these are permanently magnetised or not. The size of the magnetic field depends on the amount of current that causes it, and it alters with distance from the conductor, following an inverse-square law (Figure 6.1). Unlike electric fields, magnetic fields are of closed shapes, so that there is no start or finish. For a straight wire, for example, the shape of the magnetic field is a pattern of circles around the wire (Figure 6.2). This pattern can be detected by small compass needles, which will take up a position pointing along the direction of the magnetic field. By convention, the arrowhead of the compass needle, which normally points to the earth's (magnetic) north pole in the absence of other fields, is taken as indicating the direction of any magnetic field. We cannot use the terms 'north' and 'south' of the field around a conducting wire, because the field is a continuous pattern.

If a wire is wound into a long coil, the shape that we call a **solenoid**, the magnetic field takes a more familiar pattern when current flows in the wire (Figure 6.3). This is almost the same as the shape of the magnetic field around a permanent bar magnet, so that there is a concentration of field around each end of the

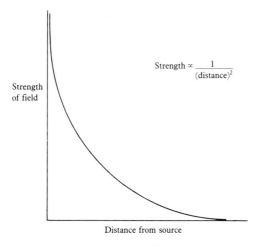

Figure 6.1 *Magnetic field strength and distance. Like most field effects, the strength of the magnetic field decreases sharply as distance from the source is increased. The relationship is an inverse-square law*

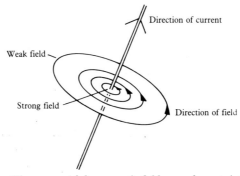

Figure 6.2 *The pattern of the magnetic field around any point on a straight wire. Reversing the direction of current through the wire will reverse the direction of the field. Each circle shown is a contour line that joins points of equal field strength*

solenoid, as there is around the poles of a bar magnet. This pattern becomes much closer to that of a permanent bar magnet if the solenoid is wound around a magnetic metal like iron. The best form of iron for this purpose is iron that has been annealed to a soft form. This is called 'soft iron', and the word 'soft' has come to have a meaning of its own when applied to magnetism. A soft magnetic material will concentrate a magnetic field, making the field much stronger, when the field is produced by a solenoid or

Magnetism 69

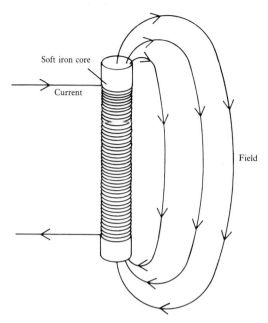

Figure 6.3 *A few lines of the magnetic field pattern around a solenoid, one side only shown. The shape of the field is the same as for a bar magnet, but the strength of field is controlled by the amount of current flowing in the coil. The lines of the pattern are most concentrated at the ends of the solenoid, which is where the magnetism is most intense*

any other source. When the source of the field is removed, as by switching the solenoid current off, the soft magnetic material is no longer magnetised. The opposite type, a 'hard' magnetic material, remains magnetised under these circumstances. The words 'hard' and 'soft' are applied in this way to materials which may all be mechanically hard, even brittle.

Force effects

When a solenoid is passing current, there will be a force on any core of magnetic material, and also on any other magnetic material, and other solenoids or any other conductors that are carrying current. The direction of force will nearly always be directed to one end of the solenoid, and most of the applications of solenoids in electronics make use of a metal piece called an **armature** which will be attracted to one end of a solenoid when current flows through the solenoid. For other purposes, the core

Figure 6.4 *An electromechanical solenoid (RS Components Ltd)*

of the solenoid is made moveable, and is pulled into the coil when the coil is energised (Figure 6.4). This use of a solenoid allows a greater range of movement than the type using an armature, and this type of solenoid action is used widely in all forms of electromechanical devices, such as vending machines, car starter mechanisms, electrically operated door bolts, valves for gases or liquids, and other industrial automation equipment. Considerable power is needed for the larger solenoids, and if the mechanism uses a spring to return the solenoid to its unenergised position, then current will have to be applied for as long as the solenoid action is needed.

For some purposes, this would require too large a continuous current, and **double-acting solenoids** are used. Current through one winding will then move the solenoid core in one direction (valve *on*, for example), and current through another winding will move the core back (valve *off*). With this type of mechanism, the windings are energised only when a change is needed.

As an example of the smaller type of solenoid used in miniature mechanisms, the RS Components 347–652 uses 12 V DC with a coil power of 3 W. This will give a force of 0.05 N over a stroke (movement) of about 4 mm, with a spring to return the core when the coil is de-energised. The force will be smaller for longer strokes up to about 12 mm maximum. The 3 W rating is continuous, and larger powers can be dissipated, with correspondingly greater thrust force, for short periods.

Another type of solenoid, the **rotary solenoid**, is constructed like an electric motor (see Chapter 7) and will revolve a shaft through an angle, typically 45 degrees, when the coil is energised. This type of mechanism is often more useful for actions like movement of tape or film, counting or sorting actions or shutter systems.

The words 'solenoid' and 'actuator' are used with almost identical meanings in electrical work to mean this type of mechanism, but in electronics, 'solenoid' is just as likely to be used simply to describe the shape of coil.

Relays

A solenoid coil shape is also used in relays, which still form an important part of electromechanical devices, though some applications now make use of purely electronic devices like thyristors (see Chapter 11). The **relay** is a type of electromechanical switch, and the term 'relay' is often reserved for the smaller types, with 'contactor' used for larger varieties. Relays exist in a baffling variety of types, but the basic principle is that current through a solenoid coil with a fixed core attracts an armature. This armature in turn operates switch contacts that make, break or changeover another circuit. The advantage of using a relay is that the circuits can be very thoroughly isolated from each other. The primary circuit is the one that operates the solenoid of the relay, and is usually a low-voltage, low-current circuit. The contacts of the relay are completely insulated from the solenoid and from the body of the relay, and can handle a range of voltages and currents according to their design. In some applications, only a mechanical relay will be acceptable to safety regulations, particularly when mains voltages are to be switched under conditions of high moisture.

The specification of a relay for a particular purpose will generally start with the switched (secondary) circuit, since this determines the contact arrangement. When the number of contacts and type (*normally open, normally closed* or *changeover*) has been specified, the primary circuit can be considered. If the secondary contacts are to carry low currents, the force needed to operate the contacts can be low, and a miniature relay can be used. The smaller the relay, the less power will be needed in the primary circuit. For some purposes, if only a very small change of voltage or current is available to operate the relay, the transistorised type of relay can be used. This will, however, need a source of power to operate the coil apart from the 'on/off' current.

Latching relays are used when only a brief pulse of current is available to switch the relay. The latching relay uses one pair of contacts to pass current through the relay coil after it has been energised for long enough to make the contacts switch over. Latching relays can be of the latch-on only type, or, more usually, can use one pulse of current to hold on, and another to hold off again. Practically all relays are available as open or closed types. The open type of relay is easier to inspect and maintain, but is more likely to suffer from corrosion and burning of contacts. The closed type can have a longer life when the atmosphere around it is corrosive, but it is then less easy to check for contact burning.

Many relays are made in more specialised form. One common type for electronics work is the **reed relay**. This consists of two or more thin metal strips, or reeds, bonded into a glass tube of small diameter (Figure 6.5). The tube can be partially evacuated to minimise the oxidising effect of the air on the contacts. The reed relay can be specified as *normally open, normally closed* or *changeover*, and can usually handle at least 100 V, and maximum currents ranging from 250 mA to 2 A, depending on contact size. The reeds in their sealed glass tubes are switched by a separate solenoid. This allows for replacement of the reeds in the event of contact failure, though for many purposes, the reed and solenoid is made as a single package. Reed relays are particularly useful if fast switching is required, because typical switchover time is 2 ms; a typical time for larger relays would be 25 ms. For applications that require low contact resistance and little or no bouncing of contacts, a mercury-wetted reed relay can be used.

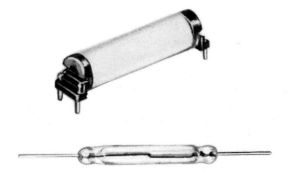

Figure 6.5 *A reed-switch and its operating coil (RS Components Ltd)*

Relays in circuit diagrams

The way in which relays are shown in circuit diagrams can often be a source of confusion. The standard system is that of *detached-contact representation*, in which the solenoid appears in its (primary) circuit, and the contacts in their (secondary) circuit, but the two need not be shown together. In some diagrams, the solenoid may be on one page of the diagram and the contacts on others. The tie-up between the two is by means of letter and number labelling, as Figure 6.6 illustrates. The box shape is used to represent the solenoid, and various symbols within the box can be used to show different varieties of mechanism, such as *slow-break*, *fast-make* and so on. This part of the relay will be marked as *RL* for relay, with an identifying letter or letters, such as *RLA*, *RLB*, and under this will be shown the number of contacts. For each contact of a given relay, the relay letter will be shown as *RLA*, *RLB* and so on, and the contact number follows. The example shows this form of marking in use to illustrate the principles.

Detached contact drawing makes it much easier to follow a circuit diagram which otherwise might have to be drawn in a very confusing form in order to bring all the connections to a relay to a single area.

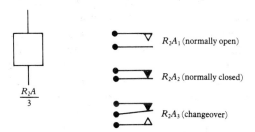

Figure 6.6 *The detached-contact system of illustrating relays on circuit diagrams. The symbol used for the coil may be modified to show the type of relay*

Magnetic materials

We have briefly touched on the idea of soft and hard magnetic materials, referring to whether magnetism is retained or not. This is just one aspect of the effect of materials on the magnetic field of a solenoid, and a full treatment of the way that materials respond to magnetism would take many books of this size. Here, then, we shall deal only with the essentials that are necessary to understand how so many electronic devices operate.

All materials are affected to some extent by the magnetic field of a solenoid or other coil shape. Most materials are classed as being either **paramagnetic** or **diamagnetic**, and the difference appears when an intense magnetic field acts on a sample of a material in the form of a rod. For paramagnetic materials, such a rod will line up with the direction of the magnetic field; for diamagnetic materials, it will turn so as to be at right angles to the direction of the field. The forces on most materials, whether paramagnetic or diamagnetic, are so weak that these effects can be demonstrated only with exceptionally large magnetic fields. There is, however, another set of materials which are so strongly paramagnetic that they form a class of their own: the *ferromagnetic* materials. The word 'ferromagnetic' simply means that these materials are magnetic like iron, because iron was one of the first materials known to behave in this way. Ferromagnetic materials are affected by magnetic fields, and also act on the magnetic fields themselves. The force between a ferromagnetic material and a magnetic field is thousands of times stronger than the force on an ordinary paramagnetic material, and the shape of the magnetic field around a ferromagnetic material is very different from that around an ordinary paramagnetic material.

These effects are related. When a ferromagnetic material is placed in the field of a solenoid, the directions of the field are completely changed (Figure 6.7). The shape of the field can be conveniently traced with compass needles, or better, by sprinkling iron filings over a sheet of paper over the solenoid and ferromagnetic bar. The patterns that are obtained show that the ferromagnetic material is able to concentrate or focus the field of the solenoid. The lines in this pattern are often called '**flux lines**,' so that we refer to the action of the ferromagnetic material as concentrating flux. The intensity of this effect can be measured by a quantity that we call **flux density**, whose symbol is B and whose units are webers per square metre.

The concentrating effect of the ferromagnetic material is sometimes compared to the effect of a conductor on electric current, so that we can refer to a magnetic circuit of ferromagne-

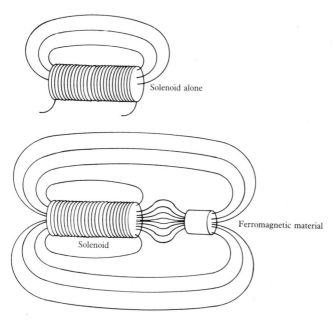

Figure 6.7 *The effect of a ferromagnetic material in the field of a solenoid (with no core). The ferromagnetic material concentrates the field as if forcing the magnetic field lines to crowd into it*

tic material which carries flux caused by the effect of a solenoid. Staying with this idea, the effect of the solenoid is measured as **magnetomotive force** (compared with electromotive force of a cell), measured in terms of the current flowing in the solenoid multiplied by the number of turns (unit ampere-turns). When we think of magnetism in terms of the circuit like this, it can often make the action of devices appear much simpler, because if any part of such a magnetic circuit can be moved, it will move so as to make the circuit complete.

The effect of a ferromagnetic material is measured by how much it can concentrate flux; this factor is known as **relative permeability**. The principle is quite a simple one. Suppose that we have a solenoid which is completely surrounded by air (or a vacuum) and has a constant current passed through its coil. This current will cause a value of flux density that can be measured at some point in the solenoid. Now if a ferromagnetic material is inserted in the solenoid, the flux density will be greatly increased, and the ratio of flux density with ferromagnetic material to flux density in air or vacuum is the relative permeability of the ferromagnetic material. Values of 20 000 or more for this quantity are common, meaning that a core of such a material will give flux density figures 20 000 (or more) times the flux density in air.

The hysteresis curve

The effect of magnetising a ferromagnetic material by passing a current through a solenoid is by no means simple. Relative permeability is not a constant quantity, and its magnitude for a given ferromagnetic material will vary according to the size of the field in which the material is placed. In addition, the way in which the relative permeability changes is different depending on whether the field is increasing or decreasing. This effect is called **magnetic hysteresis,** and it is best illustrated by considering what happens when a bar of unmagnetised ferromagnetic material is placed in a solenoid, and the current through the solenoid gradually changed. The effects are then shown by plotting the flux density of the material against the amount of ampere-turns of the solenoid. This graph is called a **hysteresis loop.**

A typical hysteresis loop is illustrated in Figure 6.8. The axes from $-H$ to $+H$ are used to indicate the number of ampere-turns of magnetisation of the solenoid, and point 0 is the zero-current mark. The axes from $+B$ to $-B$ indicate flux density in the ferromagnetic material. Plus and minus signs are used because the magnetising current can be in either of two directions, and the flux density can be in either of two directions – it is immaterial which direction we take as positive. In any graph of this type, the scale of numbers that would be used for the B line would be very large compared to the scale used for the H line. This is because the amount of flux density is always much greater than the amount of magnetic field, or ampere-turns.

If we start with no current flowing in the solenoid, the material will be unmagnetised, and conditions are represented by point 0. If we now start to pass current through the solenoid, gradually increasing the current, the amount of flux density will greatly increase. The graph is not a straight line, and where its slope is steepest represents the region in which the relative permeability value of the material is greatest. This slope is not constant,

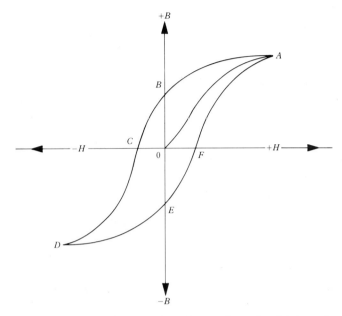

Figure 6.8 *A typical hysteresis loop shape. The section OA is used only when a material is magnetised for the first time after being demagnetised*

however, and it starts to decrease until the graph line is horizontal. This is because a material cannot be magnetised to more than a limiting value of flux density, called its **saturation flux density**. The value of saturation flux density, represented as point A on the graph, measures how strongly the material can be magnetised, and this figure is often quoted for magnetic materials. Values for a few materials are shown in Figure 6.9. This value is often useful, because many devices that rely on the action of a ferromagnetic core cannot work when the core becomes saturated. For such purposes, the materials that are used for cores will be materials that have a very large value of saturation flux density.

The curved line from 0 to A is never used again unless the material can be completely demagnetised and the measurements restarted. With the material saturated, decreasing the amount of current through the solenoid will make the graph draw out the curve through point B. This graph shape indicates that the flux density decreases only by a comparatively small amount below the

Material	Cobalt	Hypersil	Mumetal	Nickel
Saturation flux density	1.8	2.0	1.0	0.6

Note: Units of flux density are teslas (webers per square metre).

Figure 6.9 *Some figures for saturation flux density of various soft materials. These figures are approximate only and will vary from one sample to another*

saturated level, and still has a large value of flux density when the current through the solenoid is zero. This means that the material is a permanent magnet and there is flux density remaining when the magnetising field is zero. The distance shown as $0B$ on the graph represents the quantity that is called **remanence**, the amount of flux density that remains when the magnetising current is switched off. For a material that is to be used as a permanent magnet, high remanence is essential.

The next part of the curve, from B to C is traced only if the current through the solenoid can be reversed and gradually raised in the reverse direction. This part of the graph indicates that the magnetisation of the material can be reduced to zero only by applying a reverse field from the solenoid. If only a small field is needed, this indicates that the material is easily magnetised and demagnetised, and if the required field is large, the material is difficult to magnetise and demagnetise. The quantity represented by distance $0C$ is called the **coercivity** (or *coercive force*) of the material, and is used as a measure of ease of magnetisation. Figure 6.10 contains some values for this quantity, showing the large differences between material like cobalt, which is very difficult to magnetise and demagnetise, and the alloy Mumetal which is very easy to magnetise and demagnetise.

The rest of the hysteresis curve is a mirror image of the graph traced after reaching saturation. As you might expect, the same amount of saturation flux density can be achieved in the opposite direction, and there is another identical value of remanence and one of coercivity in the opposite direction.

Choice of materials

The hysteresis curve that was illustrated in Figure 6.6 was a general

Material	Alnico	Cobalt	Iron	Mumetal	Nickel	Stalloy	Steel	Ticonal
Coercivity	45000	1000	80	2.5	400	45	4300	50000

Note: Units of coercivity are ampere turns per metre.

Figure 6.10 *Some approximate figures of coercivity for various materials. The units are ampere-turns per metre, meaning current multiplied by number of turns of the solenoid winding and divided by the length of the winding in metres*

one, chosen so as to show all the possible features of such a curve. In practice, there are hundreds of magnetic materials, both pure metals and alloys, so that the range of shapes of hysteresis curves is very large. Some materials have very low remanence and coercivity. In other words, they can be very easily magnetised and demagnetised, and retain very little magnetism. Such materials are the ones that we describe as being magnetically soft. Soft materials are used for the cores of solenoid actuators and relays, among others, because we want to achieve large flux density values with very little current in the solenoid, and for the cores to lose their magnetism when the current stops. Soft materials also come in for a lot of use as magnetic shields. If electronic components (like cathode-ray tubes) have to be shielded from magnetic fields, this can be achieved by surrounding these components with soft magnetic material. The reasoning here is that the magnetic flux will be concentrated in the soft magnetic material rather than in the electronic components.

By contrast, there are few applications for hard magnetic materials in electronics apart from the permanent magnets in instruments such as moving-coil meters. Hard magnetic materials will have large values of remanence and coercivity. They will need a solenoid with a large number of ampere turns to magnetise them, but will retain a large amount of flux density, and require a large field in the opposite direction to demagnetise them.

Demagnetising materials

It is sometimes necessary to demagnetise materials that have become magnetised unintentionally. Unintentional magnetisation can come about because of a field from a coil or even from the weak field of the earth itself. In particular, if an unmagnetised magnetic material is struck while it is placed in a field, it will usually retain some magnetisation. This is because the process of magnetising a material relies on lining up groups of atoms. There is a limited number of such groups, called **domains**, and that is why materials reach saturation; when each domain has been turned along the line of a magnetic field, the flux density in the material is as high as it can be.

From what we have seen of the magnetic hysteresis curve, a material cannot be demagnetised easily. If we can use only steady currents, then a material has zero flux density only while a field in the opposite direction is applied to it. Fortunately, there are other ways. One way is heating. Each material has a critical temperature, called the **Curie point**, at which its retained magnetism will be removed. The Curie temperatures typically can be in the region of 400 °C to 1000 °C. A lot of materials cannot be heated to such temperatures in air without being oxidised, so that demagnetisation by this method is usually carried out in hydrogen gas. This type of process, also called **annealing** because it makes the materials mechanically soft, is generally reserved for the Mumetal shields that are used for cathode-ray tubes. Once such shields are annealed, they must not be drilled, riveted or hammered, because such treatments would allow them to become magnetised again. The annealing is therefore carried out only after all forming operations are complete.

The alternative to annealing is the use of AC demagnetising coils, also called **defluxing** or **degaussing coils**. The principle here is to surround the magnetised material with a coil through which current can be passed. The current that is used is AC, however, which goes through a cycle of rising to a peak value, dropping again, then reversing to an opposite peak and dropping, many times per second. The use of AC in the coil makes the material go through its hysteresis cycle in each AC cycle. If the current is slowly decreased, or the material slowly removed from the coil, the peak of the hysteresis loop will become smaller on each cycle, until its value is negligible. At this point, the material is effectively demagnetised.

7
Current and motion

The magnetic field that is created when current flows through a conductor will interact with any other magnetic field, or with a magnetic material, to produce force. If this force can be harnessed to produce rotation, we have a useful motor, a source of mechanical power. Rotation is much more useful and practical than movement in a straight line for this purpose, because it is comparatively simple to produce a magnetic field in which something can rotate, quite another thing to produce a magnetic field along a long straight line. Primitive types of electric motors were made in the early nineteenth-century, but the practicable electric motor owes its existence to the work of Michael Faraday – and the fact that no one was around to organise protest movements against progress in those days.

The motor principle

The use of the force between magnetised objects is the basis of all types of electric motors (see Figure 7.1). If we simply imagine magnetised bars, without thinking about how they come to be magnetised, you can see that the bar will be affected by the magnet, and the effect will always be to turn the bar so as to close the gaps making the magnetic circuit as complete as possible. If this were an arrangement of permanent magnets, that would be the end of it. But suppose that the bar is an electromagnet. Starting at the position with the bar at right angles to the gap in the main magnet, called the **field magnet** (Figure 7.2), imagine that the bar has been magnetised so that its magnetic polarity is as

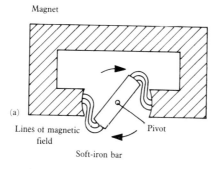

Figure 7.1 *Magnetic forces. A soft-iron bar in the field of a magnet will distort the field (a), and there will be a force on the ends of the bar pulling it into line (b)*

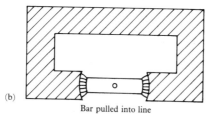

Figure 7.2 *The principle of commutation. If the magnetism of the armature is reversed just after the armature is in line with the field, then the rotation will be continuous*

shown. This will cause the bar to turn so as to end in line with the gap, but the movement is not one of turning and stopping in the final position. Any bar of metal will have some mass, and once it starts turning, it can't be stopped instantly. It will be turning at some speed as it comes into line with the gap, and will keep turning for some time after, due to rotational inertia. Imagine that as the bar turns beyond the position of being lined up with the gap, the current through the coil is reversed. Now instead of the bar being attracted back to the lined-up position, it will be forced to keep turning, since the correct lined up position is now almost 180 degrees on. If we switch the direction of the current each time the bar has just passed the lined-up position, then the bar will keep turning – and we have an electric motor.

As described, this is a primitive motor of the permanent magnet field type. The bar is called the **armature**, and the switching of current is carried out by a **commutator**, an automatic current-reversing switch. The commutator is illustrated in Figure 7.3. It consists of two conductors in the shape of split halves of a tube secured to a non-conducting shaft. These segments of tube are each connected to one end of the coil around the armature of the motor. Contact to the revolving segments is made by carbon rods, called **brushes**, which are located in tubes and held on to the commutator segments by springs. By placing the commutator segments in the correct positions relative to the position of the armature, the current through the armature coil can be changed over at the correct point in the rotation of the

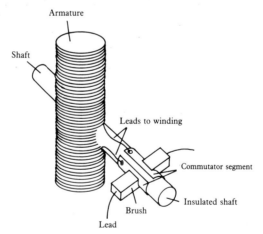

Figure 7.3 *A mechanical commutator. This consists of a split copper ring on an insulated shaft, with each segment connected to an end of the armature winding. The split is positioned so that the connections change over as the armature is in line with the field. A carbon contact, the brush, makes contact with the revolving commutator segments*

shaft. The reversal of current in this way, however, causes problems of sparking, which burns the commutator segments and causes radio interference, so that motors of this type are not necessarily ideal for electronics use.

A motor with a simple bar armature (a two-pole armature) has a very uneven rotational speed, and smoother movement is obtained if a cross-shaped armature, the four-pole armature, is used. Such an armature requires more commutator segments, four instead of two. The principle can be extended so that many poles are used in the armature, with two commutator segments to carry the current for each pole.

For all but the smallest sizes of DC motor, a wound magnet will also be used for the field. This field coil can then be connected in two basic ways, either in series with the armature coil or in parallel (**shunt**). With the field coil in series (Figure 7.4) the motor has very high turning-force (**torque**) when it is switched on, but can run at a very high speed and develop a lot of mechanical power for its size. The torque decreases as the speed increases, and the motor can reach destructively high speeds if not loaded. With a shunt field (Figure 7.5) the starting torque is fairly high, and the speed of the motor can be regulated by altering the amount of current in the field winding. This does not act in the direction that you might expect. The *greater* the amount of field current passed through the field coil, the *slower* the motor revolves, though increasing the current in the armature winding will increase the motor speed. Shunt-wound motors are useful when high starting torque is needed, along with good speed control, and a speed that remains fairly constant when the amount of mechanical load on the motor is changed. For some purposes, **compound motors** are used, with some of the field windings connected in series and some in shunt.

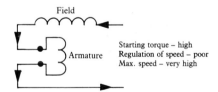

Figure 7.4 *A series-wound motor. The armature winding is connected in series with the field winding. This is a very common type of connection for small motors*

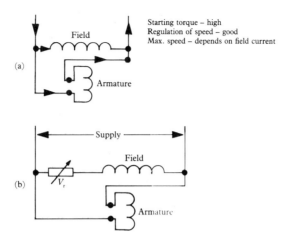

Figure 7.5 *A shunt-wound motor. (a) The armature winding is connected in parallel with the field winding. A variable resistor (b) can be connected in series with the field winding to be used as a speed regulator*

AC can be used as a supply to motors of the DC type provided that both the field and the rotor use electromagnets. For many purposes, however, AC-only motors are used, such as **synchronous** or **shaded-pole motors**. These motors use field windings in which the magnetic field rotates, and the armature consists only of a metal cylinder or a set of metal bars, which will rotate at a speed very close to the speed of rotation of the magnetic field. The synchronous motors have very low starting torque, and sometimes need to be started separately, but run at a very constant speed that depends only on the frequency of the AC supply. The shaded-pole motor uses a simpler construction, but its speed is not so constant.

The moving-coil movement

The principle of an armature coil turning so as to line up with the field of a permanent magnet is one that is used also for measurement of current. Current measurement requires the use of a field that is uniform from one position to another, and the simplest shape of field of this type to create is a radial field. This is achieved by using poles whose shape is tubular, with a rod of soft magnetic material placed centrally (Figure 7.6). The coil that acts

Current and motion 87

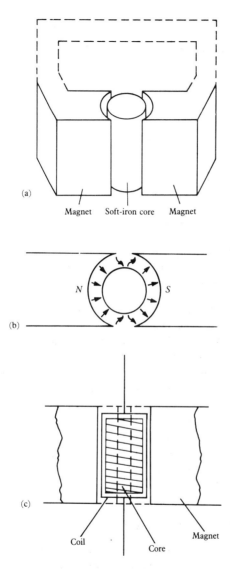

Figure 7.6 *Creating a radial field. The general arrangement (a) is of a magnet with a pair of semicircular poles and a soft-iron core placed in the gap. The field pattern (b) is radial except in the region of the open slits at each side. The coil (c) is arranged so that it turns in the space between the core and the magnet poles. In this way, the turning-force on the coil is proportional to the amount of current through the coil at all normal positions*

as an armature is then suspended so that its turns are around the central core, but not touching the core. The suspension can use springs of fine metal fibres, and these can be arranged so that the coil is placed with its sides close to the gaps in the field. When a current flows through the coil, the force between the field and the current acts to turn the coil, winding up the springs or the metal threads that hold the coil. The angle through which the coil turns is proportional to the current through the coil, so that if a needle is attached to the coil and moves over a scale, we can mark out units of current on the scale at uniform intervals. This is a linear meter movement, meaning that if the ends of the scale represent 0 and 10 current units respectively, then we can mark off the rest of the scale by dividing it up into ten uniform divisions.

The sensitivity of this meter movement is measured by the deflection of the coil (and needle) that will result from a given amount of current. One useful measure of sensitivity is **full-scale deflection** sensitivity, abbreviated to FSD. The FSD for a meter movement is the amount of current that will cause full-scale deflection, whether this is of a needle or of a light-beam reflected from a mirror carried on the coil. The smaller the amount of this FSD current, the more sensitive the meter. The sensitivity is increased by making the permanent magnetic field larger, and by making the coil larger and with more turns. These requirements are conflicting, because it's always easier to create a large magnetic field in a small space than in a large one. The compromises that are reached allow us to manufacture movements with a sensitivity of 5–20 µA with straightforward production methods.

Meter scales

A meter of about 20 µA FSD would be of rather limited use by itself, but the value of the moving-coil movement is that it can be the basis of any meter for current, voltage or resistance. Different ranges of FSD current can be obtained by connecting different values of resistance in parallel (shunt) with the meter movement. Different values of DC voltage sensitivity can be obtained by connecting suitable resistors in series with the meter movement. By using a cell and a variable resistor in series with the movement, it is also possible to measure resistor values, but not

Current and motion 89

with the convenience of the other two measurements. We'll look first at current ranges.

Suppose we have a meter movement (Figure 7.7) whose FSD current is i amperes and whose resistance is r ohms. In practice, these will be small values, like 20 µA and 1 K. Now imagine such a movement with a resistance R in parallel, shunting the movement. When a current of I amps flows, some current will pass through the meter movement and some through the shunt resistor. If we imagine that current I into the whole circuit causes FSD current i in the meter movement, then the amount of current through the shunt resistor must be $I-i$, the rest of the current.

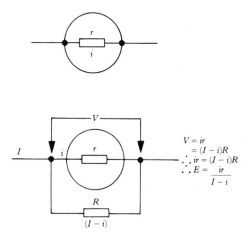

Figure 7.7 *Meter shunts. The meter has a value of internal resistance* r *and FSD current* i. *By connecting an external shunt resistor* R, *some current is diverted through this resistor and the relationship is as shown*

Now the voltage across the meter movement must be exactly the same as the voltage across the shunt resistor, because the two are connected together. This means that $r.i = R(I-i)$, and this equation can be arranged to give the equation for R shown in Figure 7.8. In words, we can arrange for a shunt resistor so that the FSD of the meter with this shunt attached is any value we please. In practice, the equation would lead to values of R which could be impossibly small, so that we often connect resistance in series with the movement before adding the shunt. The principle, however, remains the same.

$$R(I - i) = ir \qquad RI = ir + iR$$

$$R = \frac{ir}{(I - i)} \quad \text{or} \quad I = \frac{i(r + R)}{R}$$

Figure 7.8 *Rearranging the shunt equation so as to find the value of R of the shunt resistor for any desired total current I, or the value of I for any given value of R*

By using several shunts switched into circuit, we can make multimeters with several current ranges. The usual arrangement is of a **universal** shunt consisting of several high-precision wire-wound resistors in series. These resistors are connected (Figure 7.9) so that as the switch connections are changed, both the series and the shunt resistance will change, keeping the total resistance closer to a constant value.

The addition of resistance in series instead of in shunt allows us to use the movement as a voltmeter (Figure 7.10). What is actually being measured is current, but because we use ohmic

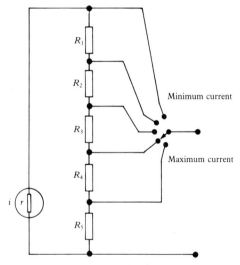

Figure 7.9 *A universal shunt circuit. In the minimum current position of the switch, the shunt consists of all of the resistors in series so that the maximum measurable current is about equal to the FSD current of the movement. In the maximum current switch position, R_5 acts as a shunt, and the other resistors are in series so that a large current is needed to give FSD*

resistors, the relationship $V = RI$ holds true, and the voltage is proportional to current. Suppose that the FSD current is i amps as before, with the resistance of the movement r ohms. By adding a series resistor of R ohms, we make the total resistance equal to $R + r$ ohms, so that the voltage across the circuit at FSD current is $i.(R + r)$ volts. This means that for any voltage V for which we want the meter to register FSD, we can calculate a resistance R to connect in series. Once again, this would be a wire-wound resistor for the greatest possible precision, and Figure 7.10 shows the method of calculating the size.

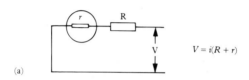

(a)

(b) Since $V = i(R + r)$
$= iR + ir$
$R = \dfrac{V - ir}{i}$
$= \dfrac{V}{i} - r$

Figure 7.10 *Using a moving-coil movement as a voltmeter. The series resistor R is chosen so that a voltage of the desired range V will cause FSD current to flow*

Resistance measurement

Resistance measurement with a meter movement is carried out with the type of circuit that is illustrated in Figure 7.11. The cell is used to pass current through a series circuit, of which one part is the unknown resistor, R. Before this is done, the terminals XY are shorted together, and the resistor VR_1 adjusted so that the meter is at full scale. This point is marked zero ohms on the resistance scale. With the resistor to be measured, R, in place, the reading of current will be lower, so that low current corresponds to high resistance. The scale is not linear, so that, for example, the 1000 ohm point is not midway between the zero and the 2000 ohm point. Several scales of resistance, each using a

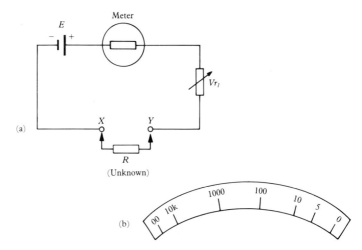

Figure 7.11 *Measuring resistance with a moving-coil movement. This requires a cell to provide an EMF, and a variable resistor to set the current to its FSD value when the terminals* XY *are shorted together. The scale of resistance (b) is not linear and reads in the opposite direction to the scales of voltage or current*

different value of VR_1, the variable resistor, are usually needed to cover a range of resistance from about 1 ohm to about 1 M.

Hall effect

The effect of a magnetic field on a current is not always to move a conductor. Current is carried by charged particles, and the force that a magnetic field exerts on a conductor is, in fact, exerted on these particles. The **Hall effect** is an example of this action, and it was the way in which hole movement was first proved. The principle is a comparatively simple one, but for most materials detecting the effect requires very precise measurements.

The principle is illustrated in Figure 7.12. If we imagine a slab of material carrying current from left to right, this current, if carried entirely by electrons, would consist of a flow of electrons from right to left. Now for a current and a field in the directions shown, the force on the conductor will be upwards, and this force is exerted on the particles that carry the current, the electrons. There should therefore be more electrons on the top surface than on the bottom surface, causing a voltage difference, the **Hall**

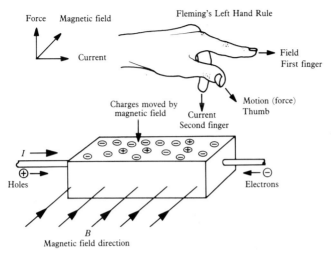

Figure 7.12 *The Hall effect. The effect of the magnetic field on the current carrier is to deflect the carrier in the direction shown. If the majority carriers are electrons, then the top face of the slab will become more negative than the lower face, giving a detectable voltage between the faces. If the majority carriers are holes, then the polarity of the voltage is reversed. The size of the voltage depends on the current used, the size of the magnetic field, and the number of carriers present per cubic metre of material. Note the use of Fleming's left-hand rule for determining the direction of force*

voltage, between the top and bottom of the slab. Since the electrons are negatively charged, the top of the slab is negative and the bottom positive. If the main carriers are holes, the voltage direction is reversed.

The Hall voltage is very small in good conductors, because the particles move so rapidly that there is not enough time to deflect a substantial number in this way unless a very large magnetic field is used. In semiconductor materials, however, the particles move slowly, and the Hall voltages can be quite substantial, enough to produce an easily measurable voltage for only small magnetic fields. Small slabs of semiconductor are used for the measurement of magnetic fields in Hall-effect fluxmeters. A constant current is passed through the slab, and the voltage between the faces is set to zero in the absence of a magnetic field. With a field present, the voltage is proportional to the size of the field, and the measurement is much simpler and more reliable than the old methods using ballistic galvanometers.

Induction

The effect of a magnetic field on a current to produce force, and so movement, is an effect that can be reversed. When a conductor moves through a magnetic field, the force on the charged particles in the conductor causes a current that will set up an EMF in a way that is very similar to Hall effect, but with the conductor itself moving. The maximum effect is achieved when the conductor is at right angles to the magnetic field and moves in a direction at right angles to both, as in Figure 7.13. The generation of an EMF in this way is called **electromagnetic induction**, and it is the basis of all the methods that we use to generate electricity from mechanical movement. As usual, it is vastly easier to make a machine that uses rotation than one that uses linear motion, so that a simple generator will use a permanent magnet to provide a field and a coil that rotates in the field (Figure 7.14). This is the basis of the type of generator that we call an **alternator**.

Experiments carried out by Michael Faraday in the early

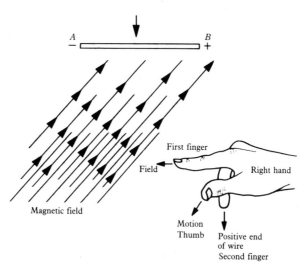

Figure 7.13 *Electromagnetic induction. A piece of wire* AB *moving through a magnetic field will cause a force on its carriers, making one end of the wire positive. Fleming's right-hand rule is a reminder of the directions of field, movement and polarity*

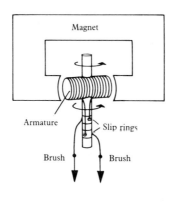

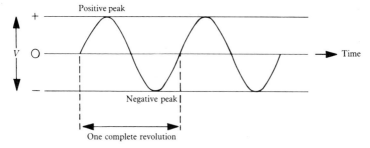

Figure 7.14 *The generator principle. The coil revolves in the field of a magnet and its ends are connected to metal rings insulated from the shaft. Brushes held against the rings provide a connection for the voltage induced in the coil, and this voltage will rise, fall and reverse as the coil revolves*

nineteenth-century showed that, for any machine of this type, the amount of EMF that was induced was proportional to the rate at which the conductors moved across the magnetic field or the rate at which the magnetic field across the conductors changed. In the simple alternator, if the coil is wound around an armature core, the field through the core will be a maximum when the core is in line with the magnet gap, and the field will be zero when the armature is at right angles to the gap. The output of the simple alternator is not therefore a constant voltage, because the rotating coil is continually changing direction. At one instant the coil will be crossing the magnetic field, but one quarter of a turn later it will be moving in line with the magnetic field, and another quarter turn later will be cutting across the field in the opposite direction.

The effect of this is that the voltage induced in the coil of a simple alternator is an alternating voltage, alternately positive and negative and passing through zero each time the field through the turns of the coil is zero. The maximum or **peak** value of EMF depends on how large the flux density of the magnetic field is, how many turns of coil are used, and how fast the armature is turned. In addition to this, the speed at which the armature is rotated affects how many complete cycles of the EMF occur per second. There will be a cycle of EMF for each revolution of the armature, so that the number of complete cycles of EMF, the **frequency** of the EMF, is equal to the rotational speed in revolutions per second.

Commutation

Commutation, as we have already seen, means the switching of connections. The output of a simple alternator is an alternating EMF, and if we need to make use of DC, then this has to be converted to an EMF which is always in one direction, even if it is not constant. Commutation can be carried out either mechanically or by using diodes. **Mechanical commutation** is carried out by commutator segments, identical to the system used for a DC motor. Commutation of this type is crude, leading to sparking and sudden changes of EMF. A much more satisfactory system for generation of DC is to commutate electronically, using **diodes** which conduct current in one direction only. An alternator designed for this purpose can be constructed 'inside-out', using a **rotor** which is magnetised with DC fed through two rotating contacts (**slip-rings**). The stationary portion (**stator**) is equipped with coils, and the rotation of the armature inside these coils will induce an alternating EMF in exactly the same way as a steady field will induce an alternating EMF in a rotating coil.

The commutation is carried out by the type of circuit shown in Figure 7.15. The diodes are semiconductor devices which pass current in one direction only, the direction shown by the arrowhead. As the diagram shows, no matter what the direction of the EMF at the inputs happens to be, the direction of current through the load is always the same because a different pair of diodes conducts for each direction of EMF. This arrangement is usually referred to as a **diode bridge**, and is represented by the

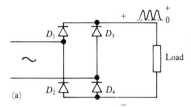

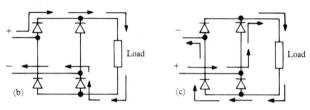

Figure 7.15 *Commutation with diodes (a). The arrowhead on the diode symbol shows the direction of flow of current. For each possible direction of input EMF, two of the four diodes will conduct and the diagrams (b) and (c) show how this leads to the current always flowing through the load in the same direction*

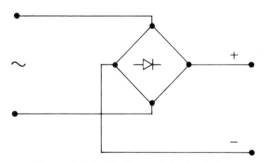

Figure 7.16 *The symbol that is often used for the diode commutation curcuit (a diode bridge) of Figure 7.15*

symbol shown in Figure 7.16. Commutation by this method involves no movement and no sparks, but is not completely perfect because the diodes do not conduct until the voltage across them is about 0.6 V. This makes the system unsuitable for very low voltages, but in all other respects it is ideally suited for power generation on a small scale and is universally used for car alternator systems. Though the output is still an EMF whose value varies between zero and peak at *twice* the frequency of the generated AC EMF, this type of unidirectional EMF is good enough for charging secondary cells, and it can be *smoothed* into true DC.

Alternating EMF

An alternating EMF is the natural form of EMF that is produced by a coil revolving in a field, or a magnet revolving inside a coil. The shape of this AC wave is referred to as a **sine wave** because the shape is exactly the same as that of a graph of the sine of an angle plotted against the angle for angles varying from zero to 360 degrees. The frequency of this sine wave is the number of complete cycles of wave per second, and the unit of frequency is the **hertz**, equal to the cycle per second. The amount of EMF at any point in the wave is called the **amplitude**, and the value of most interest is the maximum value, the **peak amplitude**. For some purposes, the value measured from one peak to the opposite peak, called the **peak-to-peak value** is useful.

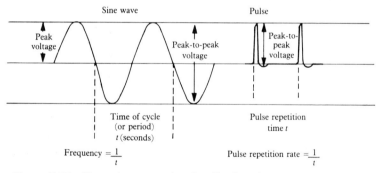

Figure 7.17 *Terms that are used to describe the voltage and timing of AC waves and pulses*

Transformers

The principle of induction is one that does not necessarily depend on the movement of a magnet. When an alternating EMF is applied to a coil, the result will be an alternating magnetic field, and this alternating magnetic field is as efficient for inducing an EMF as a changing field produced by rotating a magnet. If two coils are wound onto a single core, so that the flux of one coil will be shared by the other, then an alternating current through one coil will cause an alternating voltage to be induced in the other coil. This arrangement is called a **transformer**, and it is represented in circuit diagrams by the symbol shown in Figure 7.18b. The coil in which current flows is called the **primary** coil, and the one in which voltage is induced is the **secondary** coil.

If all the flux of the primary coil affects the secondary coil and there are no losses of any kind, a transformer will be perfect, and for such a perfect transformer, the laws of Figure 7.19 apply.

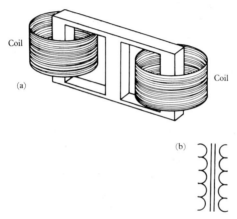

Figure 7.18 *The transformer shown in one practical form (a) and as a symbol (b). The primary winding is often wound next to the core and the secondary winding on top of this*

---

1 $V_p \times I_p = V_s \times I_s$, where V_p, I_p are primary voltage and current respectively and V_s, I_s are secondary voltage and current. In other words, no power is dissipated in a perfect transformer.

2 $\dfrac{V_s}{V_p} = \dfrac{N_s}{N_p}$ where N is the number of turns, i.e. the voltage ratio equals the turns ratio.

3 $\dfrac{I_s}{I_p} = \dfrac{N_p}{N_s}$, i.e. the current ratio is inversely equal to the turns ratio.

Figure 7.19 *The laws of a perfect transformer. Small transformers as used in electronics are likely to be far from perfect, but 90 per cent efficiency is not unusual*

Transformers are never perfect, and the smaller the transformer, in general, the lower its efficiency. One source of loss is flux leakage, that is, some of the flux from the primary coil does not affect the secondary coil. This can be minimised by careful design of the magnetic core.

Another source of loss is eddy currents. Since the core of a transformer is made of metal, EMFs will be induced in this metal as well as in the secondary winding. These EMFs will cause currents to flow, doing nothing other than heating the metal. Such currents are called **eddy currents**, and they are dealt with by making the core material as non-conductive as possible. If the core is made of metal, the metal is formed in thin laminations which are oxidised to insulate the surfaces and then clamped together. These laminated cores are magnetically equivalent to solid metal, but very poor conductors in the direction across the laminations, which is the main direction of eddy currents.

Finally, there is a loss of energy from the core material, because it is being magnetised and demagnetised in cycle. This cycle takes the core material through a hysteresis loop, and wherever there is a hysteresis loop of any type, there is a loss of energy. This cannot be eliminated, and can be reduced only by a careful choice of the material of the core and the extent to which the core is magnetised on each cycle. It is important to prevent the core from approaching **saturation**, which is why you are advised not to allow DC to flow in transformer cores if this can be avoided.

Three-phase supplies

Mains supplies are always generated as three-phase, meaning that three coils are used in the alternator and (at least) three lines used for distribution. If an alternator is constructed with three equally spaced poles at 120 degree angles (Figure 7.20), then the peak of EMF in each coil will be one third of a revolution (120 degrees) later than the coil ahead of it and 120 degrees before the coil that follows it round. This arrangement results in the generation of three waves whose peaks are always arranged one third of a cycle apart from each other (Figure 7.21). This arrangement is three-phase AC, and because the poles of the generator are arranged 120 degrees apart to produce this set of waves, we refer to the waves as being 120 degrees *out of phase*. **Phase** means a difference in timing

Current and motion   **101**

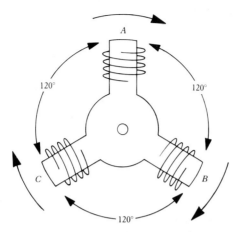

Figure 7.20 *The armature layout for a three-phase generator. Another method, used in car alternators, is to use a rotating solenoid magnetised by DC and induce three-phase AC in three stationary coils*

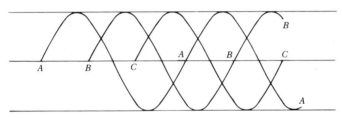

Figure 7.21 *The waves of a three-phase cycle showing the outputs from each winding*

of one wave relative to another, and it is an idea that we will come across to a much greater extent in Chapter 9. As far as electricity distribution is concerned, we can conveniently generate electricity in three-phase generators, and distribute the power along just three lines, called the **delta connection**. For distribution at low voltages, only one phase will be used, and transformers are used to convert from the delta form into the form of three live lines and a neutral, the **star connection**. A three-phase star supply can then be used for domestic distribution, using one phase line and one neutral for each of three groups of consumers. One of the main problems for the electricity suppliers is to balance the load on the three different phases so that about the same current is drawn by each phase line.

102 Electronics for Electricians and Engineers

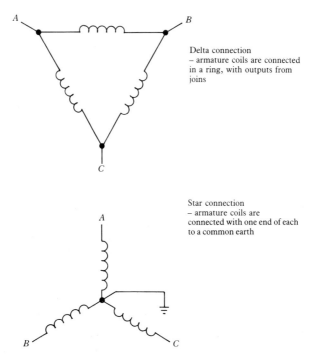

Figure 7.22 *Delta and star connection patterns in a three-phase system*

8
Ions

Most liquids are insulators, and when we consider electric current flowing through liquids, we are nearly always concerned with one particular class of liquids: solutions in water. Some molten materials, particularly metal salts, conduct, but these are of mainly specialised interest, as in the extraction of aluminium, and will not be dealt with here. In this chapter we are concerned with the conduction of current through solutions. This is possible only if the solutions contain ions, so that only solutions of this type are involved. In general, any material that is chemically an acid or an alkali and is soluble freely in water will cause ions to be formed. The materials that we call salts, crystalline materials that dissolve easily in water, will also mostly yield ionised solutions. These ions are not simply the ions of the materials itself, but often complex ions which contain strongly attached molecules of water. For more details of this process, you will need to consult a textbook of physical chemistry.

Ionised solutions

A large number of solutions of acids, alkalis, and metal salts are water-soluble and give strongly ionised solutions. The ionisation means that the solutions are conducting, so that such solutions can form part of an electric circuit. The behaviour of these solutions depends more on chemistry than on electrical laws, and depends very strongly on whether AC or DC is used. The reason is that the ions in a liquid exist *only* in the liquid. When we pass current through a liquid, we do so by immersing two metal (or other conductor) plates in the liquid and connecting these plates in a circuit. Current can then flow because the ions in the liquid move towards these plates. There will always be equal numbers of positive and negative ions, so that the positive ions will move towards the negative plate, and the negative ions to the positive plate. The ions do not generally move at identical speeds, however, so the two ions do not necessarily carry equal shares of the current.

The solution, regarded as a component in a circuit, has some unusual features. To start with, Ohm's Law is not obeyed. No current flows for small potential differences across a solution, and when current starts to flow, the amount of current is proportional to voltage only over a limited range of sizes of current. If AC is used, the solution behaves simply as a non-ohmic resistor. The situation is entirely different when DC is used, however. When DC is passed through a solution, the ions will move until they meet one of the conducting plates immersed in the solution. This complete arrangement is known as an *electrolytic cell*. At the conducting plate, or **electrode** of the cell, each ion will lose its charge, becoming a normal atom again.

When this happens, the atom will react chemically in some way, and there are three possibilities. One is that the atoms will clump together to form a material. Another is that there will be a chemical reaction between the atoms from the solution and the atoms of the electrode. The third possibility is that there will be a chemical reaction between the atoms from the solution and the water of the solution. The outcome is predictable only if the chemistry of the processes is fully understood, and in simple terms, what will happen will be whatever results in the largest possible release of energy. Very often, a combination of two or possibly all three of these effects will be found, particularly if a large current is being passed through the solution.

Electroplating

The result of passing DC through a cell containing a solution (the **electrolyte**) and two electrodes, will be the appearance of materials at the two electrodes. To take one example, if the current is passed through a solution of copper salts (such as the copper (II) sulphate), the negative electrode will be coated with copper. If the positive electrode is made of copper, this will be dissolved into the solution because of the reaction that occurs between the copper of this electrode and the negative ions in the solution. If both electrodes are of copper, copper is transferred from one electrode to the other, and the deposited copper is very pure. This type of action is called **electroplating**.

Electroplating is used to deposit thin films of a metal onto any conducting surface. In principle, electroplating requires an **anode** electrode (positive) which is made from the metal that is to be used for plating. The **cathode** (negative) is the conducting material that is to be coated. This might be another metal, but it could also be a carbon film on an insulator, or a layer of metal dust produced by the action of *sputtering* (the deposition of a metal by passing current across metal electrodes in gas at a low pressure). The last requirement is a solution of a salt of the metal that is to be deposited.

These are the fundamentals, but the technology of plating is by no means so simple. One major problem is that the metal that is to be deposited may not deposit as a polished film, but as a rough coating or, worse still, as a powder which simply falls off. Another problem is with metals, such as nickel and chromium, that have high internal stress, so that the plating deposits a thin film of the metal, but this then peels off and winds itself into a tight cylinder. These problems are overcome by choosing suitable salts in the solution and monitoring the composition very closely, by using the solution at a suitable temperature, and by careful choice of the current used for plating. In particular the current density of plating, meaning the current per square metre of surface to be coated, is important. For each metal to be used, there is usually a choice of methods that have to be adhered to very closely if good results are to be obtained.

Electroplating is one application of the use of conduction by ions in solutions. Electrolysis is also used to produce materials other than metals, for example, the gases oxygen and hydrogen and the chlorinated solutions that are used for bleaching. These applications, though important, have very little relevance to electronics. Of more importance is the use of electrolytic timers. These are miniature electrolytic cells, usually containing silver electrodes and a silver salt solution. A cell is constructed with one short electrode and one long one, and as current is passed through the cell, metal is transferred from the long electrode to the short one. This is used as a measure of time, so that the cell can be incorporated into equipment in which an overhaul is needed after some desired time of operation. By fixing the amount of current that can pass through the cell, the metal will be transferred each time the equipment is used, and inspection of the cell shows roughly how much time has elapsed. This is a low-cost alternative to the use of an electromechanical timer, and has the considerable advantage compared to digital timers that no battery backup is needed to keep the system operating when the main equipment is switched off.

Electrolytic corrosion

The action of electrolysis is not always one that is useful or desired. Wherever metals are connected to different voltages, the presence of water can cause electrolytic effects that are anything but desirable. Water is always present as humidity in the atmosphere, and this moisture can condense on cool surfaces, or be absorbed by materials such as paper, textiles or plastic foams. Pure water does not conduct to any extent, but water that condenses onto circuit boards is not likely to be pure for long. Even a minute quantity of metal or metal salt dissolving in the water can be enough to allow a few milliamps of current to flow, and once this current starts, it is likely to cause other chemical effects that will make the solution more conductive. As a result, metals can become corroded to the point where a circuit is interrupted.

Thin wires, as used in some transformers, are the most likely items to be attacked in this way, and if equipment is to be used in conditions of high humidity and high temperature, all components should be *tropicalised*, coated with resins that will prevent the films of moisture from condensing on conductors. Some components present fewer problems than others – semiconductors in general are well sealed – but wherever a metal case is used, there is a chance that it can be corroded until the metal has been completely removed at some place. No component can be completely immune unless it is encapsulated in inert resins.

The most difficult problem is the specification of electronic equipment for use at sea. Sea water is already a reasonably good conductor, and the presence of common salt (sodium chloride) in solution means that any electrolytic effects will release acidic solutions of chlorine which are very corrosive. The design of high-voltage equipment for use at sea is therefore a very specialised matter, and not one that should be undertaken lightly. Equipment for the navy is subject to very precise specifications and component approval, and any firm that tenders for contracts for such equipment must be aware of the problems that are involved.

Anodic protection

Anodic protection is the counterpart to electrolytic corrosion, a method of protecting metal that is to be immersed in salt water. The principle is to form a cell of which the salt water is the electrolyte, the metal to be protected is the cathode, and a block of otherwise useless metal is the anode. This is useful only if the metal chosen for the anode is more chemically active than the metal to be protected, and a favourite for this purpose is magnesium. A slab of magnesium can be connected to the hull of a vessel, and will be attacked by the salt water, causing a current to flow between the magnesium and the metal of the ship's hull. The direction of this current is such that magnesium is dissolved, but the metal of the hull is not, so providing protection against corrosion for as long as the magnesium exists. The magnesium **sacrificial anode** must be renewed at intervals.

Ionised gases

Up until now, we have been dealing with ions in liquids that are solutions of metal salts. Gases, such as the nitrogen and oxygen mixture that we call air, are normally non-conducting with no trace of ions. Ions can be generated in gases by a number of ways, but at normal pressures (comparable with the pressure of the air around us), these ions recombine very rapidly. Ionisation can be caused by the presence of high voltages (round power lines, for example), by the action of intense ultra-violet light (in the outer atmosphere) or by bombardment with particles like electrons from disintegrating atoms (the principle of detecting radioactivity). Whatever the cause of ionisation in air at normal pressure, the ions recombine, and affect nothing that is more than a few millimetres from where they are created. Ionisation of gases is of more practical interest when the gas that is ionised happens to be at a low pressure.

The reason is an effect called a **chain reaction**. Suppose that an electric field exists across a gas because of two electrodes at different voltages. Now if a pair of ions happens to be generated in the gas – remembering that ions must *always* exist in pairs – then the ions will be affected by the field. The negative ion will be attracted to the positive electrode and the positive ion to the negative electrode. If the ions move, however, they will collide with atoms of the gas, losing energy and travelling only a very infinitesimal distance, the average distance between atoms. Since this distance is so small, the ions will normally recombine after only a few collisions.

Conditions are very different if the gas pressure is low. A low gas pressure means that the atoms are more widely separated. The separation is still only a few millionths of a millimetre, but this can be enough to allow the ion some space in which to accelerate before striking an atom. If the ion can reach a high enough speed, then it will have so much energy when it strikes an atom that the atom will be split into ions in turn. These ions will then be accelerated, and will create new ions, and so on. The effect is that the whole of the gas will become ionised in a very short time, a matter of a microsecond or so. This ionised gas is a good conductor, and if current flows through the gas, the ionisation will be maintained because any ions that recombine are likely to be separated again by collisions with other ions.

Any device that is to make use of gas ionisation, therefore, should use gas at low pressure (less than 1/100 of normal atmospheric pressure) contained between electrodes to which can be applied a voltage high enough to start and maintain ionisation. The voltage that is required depends on the gas pressure and on the background radioactivity count. At any part of the Earth where old granite rocks are exposed, the radioactive background count will be high, and there can be quite substantial counts in houses made of deep-mined rock, bricks from deep-mined clay (or burning coal from deep mines). These counts are, in fact, higher than those from low-level waste from nuclear power-stations, but there is no political gain to be made by protesting about rocks, clay or coal. This background level of radioactivity contributes to ionisation, as does the radiation of particles from the sun, which is a large nuclear furnace. The voltage across the electrodes in a gas that is needed to make the gas conduct will therefore vary from day to day and place to place, so that all devices that make use of ionised gas have to make some allowance for this variation. The temperature also has an effect, as you will have noticed if you have had problems with fluorescent lights in cold rooms.

One of the main applications of ionised gas is, as I have hinted, in lighting. When the ions of a gas recombine into atoms, energy is given out. Instead of this being given out as electrical energy (the cause of the ionisation), it is given out in the form of light. When a current is passed through a gas at low-pressure, then, the gas glows because of the recombination of ions. There is a balance between ion formation and ion recombination, and this balance will alter as the current changes. The most direct use of this effect is in **discharge lighting**, the correct name for the type of light that is called **neon lighting**. The name comes from the first type of discharge lighting that used neon gas in a tube fitted with electrodes at each end. Discharge lights can use a range of gases and even metal vapours (as in street lights), and all of them require high voltages to start and maintain ionisation, because the gases or metal vapours are used at low pressures. In some types, heating coils are needed for starting to ensure that the pressure will be high enough to allow the light to start working.

Discharge lamps are for specialised purposes, but **fluorescent lighting** is universally used. The fluorescent light is a combination of a mercury vapour discharge lamp and a set of powders that act as phosphors. In this sense, a phosphor is a material that emits light when it is bombarded by particles or by light of a different wavelength. A mercury-vapour discharge is easy to start and maintain, but the 'light' from it is mainly invisible ultra-violet which by itself is damaging to the eyes. The fluorescent tube is coated on its inside with phosphor powders that convert the ultra-violet from a mercury-vapour discharge into visible light. The light that is given out from the fluorescent tube comes almost entirely from the phosphor, and its colour and intensity depends very greatly on the type of phosphor material that is used. Like any form of discharge lamp, special arrangements have to be made for starting.

Figure 8.1 shows the typical **switch-start** arrangement. The tube is fed through an inductor connected in an AC circuit. The inductor serves to limit the amount of current that flows, because the resistance of the ionised gas in the tube is low, and this resistance decreases as the current increases. The starter switch is a device that connects the filaments at each end of the tube so that they heat when the tube is switched on. This heat vaporises some mercury (a liquid metal), and when the switch contacts break, there is a surge voltage in the inductor which will ionise the vapour, starting the gas discharge. The starter switch can use either a bimetallic strip or another gas-filled device. An alternative method, which gives quicker starting free of the flickering that is typical of the switch-start type, is the **transformer start**. In this type, the filaments are heated by a transformer which also applies the higher voltage across the tube until the vapour ionises. The fluorescent principle is important, because it is the basis of many high-efficiency lights, some of which make use of electronic controllers (Figure 8.2) for starting and for controlling the running of the light.

Thyratrons

An application of ionised gas which is much more closely linked to electronics is the **thyratron**. A thyratron is a form of switch in which a gas acts as the switch contacts. To switch the thyratron

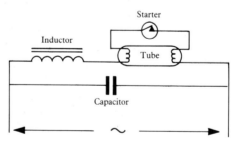

Figure 8.1 *The fluorescent tube circuit. The mercury in the tube must be vaporised by the heaters before the discharge can start. When the starter switch (a bimetallic strip) opens, the surge of voltage across the inductor starts the discharge and the tube glows, with the current limited by the inductor. The capacitor is used to correct the phase difference between current and voltage caused by the inductor*

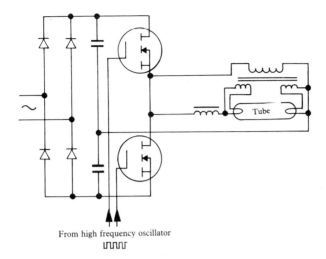

Figure 8.2 *A typical electronic controller circuit for a fluorescent tube, using power MOSFETs. This circuit uses the AC supply to provide DC that is then converted to high-frequency square waves. No starter switch is needed, starting is immediate and efficiency is improved*

on, the gas is ionised so that a large current can pass through the gas by way of two electrodes, the anode and the cathode. The thyratron is switched off when the current between the electrodes drops to a value too low to sustain ionisation, or when the voltage reverses. Thyratrons are used in applications that require a very small power to control a very large current at high voltages, possibly several hundred amps at several hundred volts.

The ionisation of the gas is achieved by a **trigger** electrode. This consists of a pointed spike close to the cathode of the thyratron. The gas in the thyratron (often hydrogen) is maintained at a constant low pressure, and because the trigger electrode is so close to the cathode, only a comparatively small voltage is needed between these two to start ionisation. Once ionisation is triggered, it will be maintained by the main discharge. If the thyratron is used in an AC circuit, it will conduct only for the positive half-cycles of the wave, and for only as long as the trigger electrode is activated.

By using thyratrons instead of diodes in bridge circuits (see Chapter 7), unidirectional current can be obtained and controlled. Thyratrons are used in welding controllers and in the control of very large electric motors. Some of the applications of the smaller thyratrons have now been taken over by the semiconductor equivalent, the thyristor (see Chapter 11).

9
AC circuits

An alternating current is the result of using an alternating EMF as a source. As we have seen in Chapter 7, an alternating EMF is the natural result of generating electricity from any rotating generator in which a magnet and coils are used. The feature of such a generator, usually referred to as an **alternator**, is that the EMF varies in the form of a wave. This means that a graph of EMF plotted against time (Figure 9.1) has the shape of a wave, and for mechanical generators will take the shape of a sine wave. This is reflected in the formula for the size (amplitude) of the EMF at any instant in the wave, which is shown in Figure 9.2. The quantity E_0 is called the **peak amplitude** and the quantity f is called the **frequency** of the wave.

The frequency of a wave that has been generated by an alternator is related to the speed of rotation, and the standard frequency for mains supplies in the UK is 50 Hz, meaning 50 complete cycles per second. The US standard is 60 Hz, and this frequency is used in the American continent and in the Far East.

The peak amplitude and the frequency are the two important measurements relating to the AC wave. When an AC EMF is applied to a circuit, there will be a current wave which will have the same waveform, but in this case, the peak amplitude will be a current amplitude, measured in amps. For a circuit that is purely resistive (containing ohmic resistors only), the relationship $V = RI$ will hold good, provided that the measurements are made on comparable quantities. This might mean using the peak values of both EMF and current, but is more likely to refer to RMS quantities.

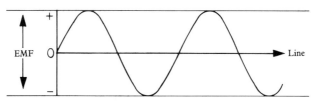

Figure 9.1 *The form of a sine wave. The shape of the wave is identical to the shape of a graph of the sine of an angle plotted against angle*

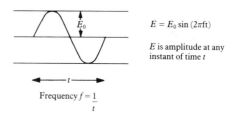

$E = E_0 \sin(2\pi ft)$

E is amplitude at any instant of time t

Frequency $f = \dfrac{1}{t}$

Figure 9.2 *The peak amplitude, time and frequency for a wave, and the equation for amplitude at any time*

Root mean square

Using peak values of AC wave quantities is unusual in power engineering, though common for electronics purposes. When we make use of mains AC, or other supplies that use sine waveforms, it is more usual to work with a calculated quantity called the **RMS value**, for either voltage or current. The reason is that when peak voltage across a resistor is multiplied by peak current, the result is not the true dissipated power, but a value that is twice as much. The reason is obvious – the peak values are not the same as average values. A simple average value, however, is always zero, because the positive peak of the wave is equal to the negative peak. By making use of the power, however, we can calculate the root mean square, i.e. the square root of the average of the square of each quantity. For a sine wave, as Figure 9.3 shows, this amounts to using the peak values divided by the square root of 2. If these RMS quantities are used, then AC circuits can be treated like DC circuits as far as power dissipation is concerned, using the RMS values as if they were DC values.

The important points to note are that these RMS values apply to sine-wave shapes only, and that they are used mainly in power

> Voltage $\bar{V}_{peak}$ Current $\bar{I}_{peak}$ Resistance R
>
> For a sinewave: Power $= \dfrac{\bar{V} \times \bar{I}}{2}$ or $\dfrac{\bar{V}^2}{2R}$ or $\dfrac{\bar{I}^2 R}{2}$
>
> Alternatively, if $V_{rms} = \dfrac{\bar{V}}{\sqrt{2}}$ and $I_{rms} = \dfrac{\bar{I}}{\sqrt{2}}$
>
> then Power $= V_{rms} \times I_{rms}$ or $\dfrac{V_{rms}^2}{R}$ or $I_{rms}^2 . R$

Figure 9.3 *Power related to AC voltage and current. The peak values can be used if all expressions for power are divided by 2. It is usually easier to work with the RMS values which can then be used just as we use DC values. Most AC meters are scaled in terms of RMS values*

engineering. For many purposes in electronics, we make use of peak values, though when power levels are being calculated, we need to use RMS values. The conversions are shown in Figure 9.3, but for sine-wave shapes only. Any meters that you use for AC voltage and current will normally be calibrated in terms of RMS quantities, and these also apply only to sine waves. The reason is that other waveshapes have a different relationship between peak values and RMS values.

Phase shifts

As far as circuits that contain only resistance are concerned, an AC circuit using a sine-wave supply can be treated just like a DC circuit, with RMS values used in any calculations. The $V = RI$ relationship holds good, and power is calculated in any of the usual ways, using RMS values. Things are very different when the AC circuit contains a capacitor or an inductor or both, because these components create a phase difference. A phase difference implies that the peak of the current wave does not coincide with the peak of the voltage wave, and for circuits that consist only of a capacitor or an inductor, the difference between the waves is one quarter of a cycle. This corresponds to 90 degrees of revolution of a generator, and this is the most common way of expressing the phase difference (Figure 9.4).

Phase differences can be explained by what happens in the components. Taking a capacitor first, suppose that it is connected

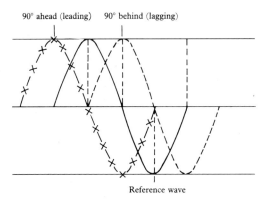

Figure 9.4 *Phase lead and lag by 90°, one quarter of a cycle. This is the amount of phase difference between the waves of current and of voltage that will be caused by a reactive component (capacitor or inductor)*

to an AC supply at a time when the EMF is passing through zero (Figure 9.5). As the voltage across the capacitor rises, the capacitor will charge, and the charging rate is fastest when the capacitor is completely discharged, that is when the voltage is about zero. By contrast, the capacitor will not be charging at all when the voltage has reached its peak value, because a capacitor charges only while voltage is changing. This means that the rate of flow of charge, which is what we call current, is greatest when

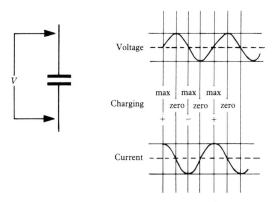

Figure 9.5 *Charging and discharging a capacitor by an alternating voltage. The charging rate is greatest when the voltage is changing rapidly, passing through zero. The charging is zero when the voltage is maximium, positive or negative. This leads to the current wave being 90° ahead of the voltage wave*

voltage is zero, and the flow of charge is zero when the voltage is a maximum. When the voltage starts to fall, the capacitor will discharge, and the same reasoning applies – the rate of discharging is greatest as the voltage is changing fastest, passing through the zero mark. The rate of discharging is zero when the voltage reaches its negative peak. The result of all this is that the current peak comes one quarter of a cycle ahead of the voltage peak. This can be written as the current peak being 90 degrees ahead of voltage, or voltage being 90 degrees behind current. We can also say that current **leads** voltage by 90 degrees or that voltage **lags** current by 90 degrees. Whichever way is used to express the effect, the idea is the same – one quarter of a cycle difference in the timing.

The same reasoning can be applied to a circuit that contains an inductor. In this case, there is an EMF generated across the inductor when the current is changing, and this EMF opposes the voltage that is applied in the circuit. Because the reverse EMF is maximum when the current is zero, and the EMF is zero when the current is maximum, there will be a quarter-cycle difference once again. In this case, however, the AC voltage peak measured across the inductor leads the current, coming one quarter of a cycle before the peak of current, as Figure 9.6 shows.

A theoretically perfect capacitor would cause a phase angle between voltage and current of exactly 90 degrees, as would a theoretically perfect inductor. Apart from electrolytics, capacitors generally are almost perfect in the sense that the phase angle caused by a capacitor in a circuit is almost exactly 90 degrees.

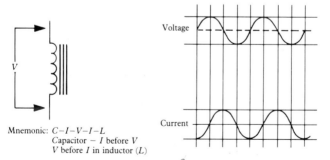

Mnemonic: $C-I-V-I-L$
Capacitor – I before V
V before I in inductor (L)

Figure 9.6 *Phase relationships between voltage and current in an inductor, and a method for remembering the phase relationships*

Inductors, however, have to be made from coils of wire which will have resistance, and this resistance causes the phase angle between voltage and current to be less than 90 degrees. For many purposes, however, we can treat inductors as if they were perfect in this sense. We shall look at some of the consequences of the phase difference later in this chapter, but one important point for the moment is that there can be no power dissipation in a perfect capacitor or inductor. Since the voltage is zero during the current peak and the current is zero during the voltage peak, the product of current and voltage averages out at zero over a cycle. Any dissipation in a circuit that contains capacitance or inductance is due only to the resistance that exists in the circuit. Such resistance will be due to the resistance of the wire of an inductor, or to an imperfect (slightly conducting) material used between the plates of a capacitor.

We have noted so far that the effect of a capacitor or an inductor is to make the phase between current and voltage (ideally) equal to 90 degrees. There is also an effect on signal amplitude. Just as we can write $V = RI$ for the effect of a resistor in a circuit, we can write $V = XI$ for the effect of a capacitor or inductor. The quantity X is called **reactance**, and unlike resistance, this is a quantity that is variable. The reactance of either a capacitor or an inductor depends on the size of the component (capacitance in farads or inductance in henries) and also on the frequency of the signal that is being used. The effect of frequency is entirely different for these two components, as Figure 9.7 illustrates. The reactance of an inductor is zero when the frequency is zero, and it rises as the frequency rises. For a capacitor, the reactance is infinitely high at zero frequency (DC), and it drops as the frequency is increased. The reactance is measured in ohms, because it is a ratio of volts to amps like resistance, but there is no other resemblance because reactance applies only to AC, not to DC. As far as DC is concerned, an inductor is a very low resistance, and a capacitor is an open circuit. For AC, both of these components have values of reactance that depend on the component size and the frequency of the AC in the circuit.

Many of the applications of capacitors and inductors in circuits are simply as separators of AC from DC, or the separation of AC signals of very different frequencies. These circuits depend on the

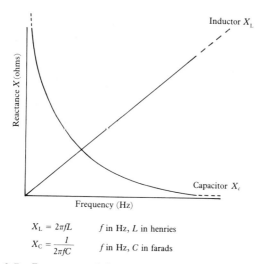

$X_L = 2\pi f L$ f in Hz, L in henries

$X_C = \dfrac{1}{2\pi f C}$ f in Hz, C in farads

Figure 9.7 *Reactance and frequency for capacitors and inductors*

difference between reactance and resistance values, or on the difference between reactance values for different frequencies of waves. Circuits that are designed to separate signals whose frequencies are fairly close to each other are called **filters**, and their design requires a very extensive knowledge of wave and electrical theory.

Many circuits make use of both a reactance and a resistance in a circuit to which a wave will be applied. The total effect of a reactance and a resistance cannot be found simply by adding the values, because in the reactance, the voltage is not in phase with the current. The combination of a reactance and a resistance is called an **impedance** and Figure 9.8 shows how the value of an impedance is calculated. The word impedance is used very much

$$Z = \sqrt{R^2 + X_C^2} \qquad \phi \times \tan^{-1}\left(\dfrac{X_c}{R}\right)$$

$$Z = \sqrt{R^2 + X_L^2} \qquad \phi \times \tan^{-1}\left(\dfrac{X_L}{R}\right)$$

Figure 9.8 *Calculating the size of an impedance when a reactance and a resistor are in series. The phase angle ϕ can also be found for these simple cases*

120 Electronics for Electricians and Engineers

in electronics, often to describe the combination of resistance and capacitance, but very often the resistance is the dominant part of the impedance, so that we can work with input and output resistance rather than impedance.

Oscilloscopes and signal generators

For electronics purposes, the sine wave is a type of signal that is very seldom used, though it still has very great theoretical importance. This is because any other waveform can be thought of as being due to a mixture of sine waves of different amplitudes and phases, but of harmonically related frequencies. The phase 'harmonically related' means that there will be one frequency which is the frequency of repetition of the wave and which is called the **fundamental**. The other frequencies that make up the wave are called **harmonics** (Figure 9.9), and their frequencies will be equal to the fundamental frequency multiplied by a whole number. The second harmonic, for example, is the frequency that

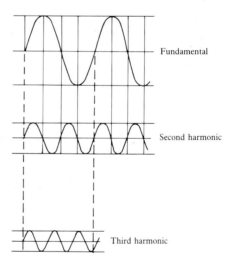

Figure 9.9 *Fundamental and harmonics. Harmonics are waves whose frequencies are some whole-number multiple of the fundamental. A wave of any shape can be analysed as a mixture of sine waves whose fundamental frequency is the same as that of the other wave. For example, a square wave can be analysed as a sine wave fundamental and a mixture of all the odd-numbered harmonics (3rd, 5th, 7th, 9th, etc.) at smaller amplitudes*

is twice the fundamental frequency, and the third harmonic is the frequency that is three times the fundamental frequency. We normally generate and measure waves, however, in ways that do not depend on mixing sine waves together.

The measurement of electrical waves is done almost exclusively by the use of the **oscilloscope**, whose full name is cathode-ray tube oscilloscope, abbreviated to **CRO**. This instrument (see Chapter 10) will produce, on a screen, a trace of a waveform for the signal connected to its input. If the screen is suitably calibrated, it is possible to measure the voltage that corresponds to signal amplitude, and the time that corresponds to the time of one cycle (Figure 9.10). The amplitude that is measured is usually peak-to-peak, because this is a more useful quantity for the type of waveshapes that are encountered in electronics. The time of one cycle can be used directly, or it can be inverted to give the fundamental frequency for the wave. For any waveshape that is not a sine wave or which is a sine wave outside the frequency range of 40–100 Hz, the use of the oscilloscope is almost essential for measuring purposes. This is because meters generally do not give correct readings for waves other than sine waves or outside the frequency range of power supplies, though some meters can give reasonable results for frequencies up to several kHz. A few modern electronic instruments can also measure the true RMS value for waveforms that are not sine waves.

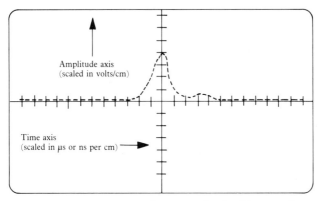

Figure 9.10 *Measuring a waveform, a pulse in this example, on the oscilloscope screen. The quantities measured are peak-to-peak amplitude and the total time of the pulse. If two or more pulses are displayed, the time between them can be measured and the frequency calculated*

The waveforms that are used in electronics are not generated by rotating machines, but by **signal generators**. Signal generators (or waveform generators) are electronic instruments that can supply a variety of waveforms (sine and square at least, often triangular and pulse as well) over a range of frequencies. Signal generators are essential for testing purposes if circuits that are being tested do not contain any source of signals. In addition, signal generators may be needed to provide calibrated signals for any circuits that carry out measurement or for which signals need to be of a specified amplitude or frequency. The combination of a good wide-range signal generator and an oscilloscope provide for most of the test and servicing work that is needed for all but some specialised electronics circuits.

Radio waves

We came across, earlier, the idea that the combination of an alternating electric field and an alternating magnetic field constituted a radio wave. The radio waves that we deliberately generate and use all start as sine waves, though they are generated electronically rather than by rotating machinery. The frequency of such waves, called carrier waves, can range from around 100 kHz, a very low (long wave) frequency, to well above the 1000 MHz (microwave) range. Radio carrier waves are classed either by frequency or by wavelength (meaning wavelength in air), and the relationship between these two is illustrated in Figure 9.11. A circuit that generates such waves can be connected to an aerial, which is simply a piece of wire whose dimensions are some suitable fraction of the wavelength, and which will act as a connection between the waves in a cable and the waves in air or

$$\text{Frequency} = \frac{300}{\text{Wavelength}}$$
or
$$\text{Wavelength} = \frac{300}{\text{Frequency}}$$

Wavelength in *metres*

Frequency in MHz

Figure 9.11 *Frequency and wavelength relationship. The units are the most convenient ones of MHz and metres*

space. An aerial that is suitable for transmitting waves will also be suitable for receiving waves of the same frequency, though reception is usually possible with aerials that are not so strictly of the correct dimensions.

Simply transmitting a sine wave and receiving it is not of much use in communication, and we have to modulate the wave. **Modulation** means changing the wave in such a way as to communicate information. The earliest modulation consisted of switching the carrier wave on and off in the pattern of Morse Code. A later form of modulation that is still used is **amplitude modulation**, and consists of changing the amplitude of the carrier so as to give the pattern of some wave signal. This is illustrated in Figure 9.12 to show how a square wave signal can be used to modulate a carrier. The point of this is that the square wave signal might be at too low a frequency to transmit, but the modulated carrier is at a frequency that can be easily transmitted. Furthermore, because the carrier frequency can be held constant, a receiver can be made to select this frequency so as to receive this signal and no others. This principle is called **tuning**.

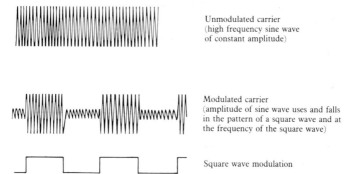

Figure 9.12 *Amplitude modulation of a carrier by a square wave. In practice, the modulation waveform would be of a more complicated shape of speech or video signal. We can also use frequency modulation in which the frequency of the carrier is varied, keeping the amplitude constant*

Tuned circuits

To understand tuning, we have to look again at the effect of capacitors and inductors on AC signals. Suppose we have an inductor and a capacitor connected in series (Figure 9.13). Given

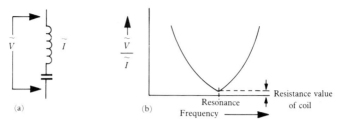

Figure 9.13 *The behaviour of a series L-C (inductor-capacitor) circuit (a) as the frequency of a signal across it is changed. The reactance V/I (b) decreases to a minimum at the resonant frequency. The value of V/I at this frequency corresponds to the resistance of the circuit, and the current is a maximum. There is no phase difference between voltage and current at resonance, and the voltage across either the capacitor or the inductor is a maximum*

the shape of the graph of Figure 9.7, you can see that there will be some frequency, called the **resonant frequency**, at which the two reactances are exactly equal. The phase shifts that the components cause, however, are opposite in direction. If we think of a current flowing at this resonant frequency, it will cause a voltage across the inductor which will be exactly equal and opposite to the voltage across the capacitor. In the circuit as a whole, then, there will be zero voltage, because the voltages across the components cancel. In other words, the series combination of capacitor and inductor has no reactance at the frequency of resonance, so that a huge amount of current can flow. In practice, there will be a small amount of resistance, but the principle is unchanged – this circuit will pass an alternating current very easily when the frequency is the resonant frequency, and signals at all other frequencies will not pass so easily. The opposite is true of a parallel circuit of capacitance and inductance (Figure 9.14). A circuit like this will behave like a very large resistance for a voltage at the resonant frequency, but a low value of reactance for other frequencies. These circuits therefore allow us to select signals of specific frequencies, and the formulae show how the frequency of resonance is related to the values of capacitance and inductance of the components.

Resonance allows us to tune radio signals. If, for example, we connect an aerial to a tuned circuit as shown in Figure 9.15, then there will be a fairly large signal across the resonant (tuned) circuit at the frequency of resonance, but not at any other

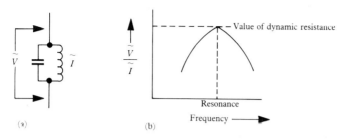

Figure 9.14 *The parallel-resonant circuit (a). The behaviour shows a peak value of V/I which is resistive (no phase difference between V and I). This quantity is the dynamic resistance, and its ohmic value is greater than the reactance of either component at the resonant frequency*

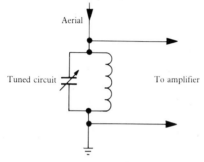

Figure 9.15 *The basis of radio reception. A tuned circuit connected between aerial and earth will provide maximum signal at the frequency of resonance. If this frequency can be varied, different signals can be tuned in*

frequency. If this is used as the input to a radio receiver, then, it makes possible the selection or tuning of wanted signals. As it happens, we use more than one stage of selection, but the principles remain unchanged.

Selection is never perfect, and a tuned circuit will not pass just one frequency but a range of frequencies around the selected frequency. This range is measured by the bandwidth, and the bandwidth of a tuned circuit is expressed as the frequency difference between the tuned frequency and the frequency whose amplitude is about 70 per cent of the tuned frequency (Figure 9.16). The choice of 70 per cent may seem odd, but it corresponds to a signal which is of half the power of the wanted signal, since 70 per cent of voltage multiplied by 70 per cent of current amounts to 49 per cent of power.

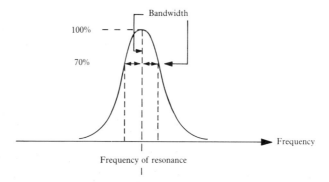

Figure 9.16 *Bandwidth is taken as the range of frequencies between the frequency of resonance and either of the two frequencies at which the signal amplitude is 70 per cent of peak value (70.7 per cent to be exact)*

The bandwidth of a tuned circuit can be decreased by making the resistance of the inductor very low, and increased by adding resistance in parallel. For some types of signal, we need the bandwidth to be greater than the tuned circuit would normally supply, and we have to connect resistors, called **damping resistors**, in parallel with the tuned circuit in order to increase bandwidth. The alternative method is to use several stages of tuning, with each stage tuned to a slightly different frequency. This is called **stagger tuning**. Where very narrow bandwidths are needed, tuning with inductors and capacitors becomes impractical, and components such as crystals have to be used instead.

The superhet receiver

The type of receiver for radio waves that is almost universally used is the superhet, an abbreviation for **supersonic heterodyne**. The reason for the name goes back to the early days of radio; each incoming radio signal has its frequency changed by an action called heterodyning, into an intermediate frequency which is higher than the frequency range of sound, and therefore called supersonic. The reason for the method lies in the problems that arise if any attempt is made to amplify weak radio signals to an amplitude which is sufficient to become useable. If you have an amplifier which, for the sake of argument, will amplify one thousand fold (has a **gain** of one thousand), then it becomes very difficult to keep this stable. If just one thousandth of the output of the amplifier, for example, should find its way back to the input and be in the same phase as the input, then the amplifer can provide its own input. It will, in other words, oscillate and generate a signal irrespective of any signal coming in. This is an effect of **positive feedback**, the transfer of a fraction of the output back to the input in the correct phase to cause further amplification. Feedback like this can take place because of stray capacitance, transformer action between two wires, of just the radiation of waves from the output so as to reach the input. Positive feedback is so difficult to prevent that it makes it almost impossible to design any amplifier with very large gain (**amplification**) that will allow the frequency of tuning to be varied. Large gain *can* be achieved if the frequency is fixed, so that the amplifier input and output can be shielded really effectively.

The superhet principle is illustrated in Figure 9.17. Each signal is tuned at the input of the receiver, using a resonant circuit in which the capacitor or the inductor can be changed in value so as to select the desired frequency. The selected range of frequencies is then mixed with a sine wave generated by a **local oscillator**. This consists of a circuit that will generate a sine wave whose frequency is *also* varied by the tuning controls. Now when two different frequencies of signals are mixed together in a suitable

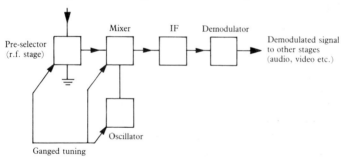

Figure 9.17 *The superhet principle. The RF and mixer stages are tuned to the frequency of the incoming signal, but the oscillator is tuned to a higher frequency. The difference between these frequencies is the IF signal which will carry the same modulation as the input signal, but which is of the same fixed frequency no matter what input frequency is selected, because of the ganging of the tuning in the mixer and the oscillator stages. The RF stage is omitted for AM radios, and some FM radios use an untuned RF stage, or one that is of broad bandwidth*

circuit, the result is a set of frequencies that includes the difference frequency. This means a signal whose frequency is equal to the difference between the incoming frequencies, but which carries any modulation of both frequencies. Since the oscillator frequency is a sine wave, the modulation of the difference frequency is the same as the modulation of the incoming signal. The difference frequency produced in this stage is called the **intermediate frequency** (IF).

The important point about the design of the superhet receiver is that the tuning of the input and the tuning of the oscillator are controlled together, or **ganged**. This is done in such a way that the intermediate frequency is always the same. For AM sound receivers, the standard IF is at 465 kHz, for FM receivers, 10.7MHz is used, and for TV, a range of IF from 33 to 38 MHz is used. Because the IF is fixed, it can be amplified in a set of amplifiers that can be screened from each other, preventing positive feedback. The IF amplifier stages can also be designed to produce whatever bandwidth is needed. This will be very small if the receiver is used for voice communication only, wider for music, wider still for high quality FM transmissions, and very wide for TV and radar use. When the IF signal is demodulated, the original signal can be recovered and because of the large gain that can be used in the IF stages, this signal can be at a fairly high level, meaning that its amplitude can be large enough to avoid the need for much more amplification.

Time constants

The use of capacitors and inductors in tuned circuits is particularly applicable to radio technology, but a very large part of modern electronics is concerned with **digital signals**. These are signals that have square waveforms, with sharply rising and falling voltages. Such signals cannot be dealt with by using the type of circuits that are used for radio transmitters or receivers, and we need to be able to predict what the effect of such signals will be on circuits. As it happens, inductors are seldom used in such circuits, so that if we understand the effect of circuits containing capacitors and resistors only we have enough information to deal with such circuits. The important fundamental idea here is the charging and discharging of a capacitor through a resistor.

Suppose we imagine a capacitor connected in series with a resistor as shown in Figure 9.18. The capacitor is discharged and the switch is open, so that the voltage across the capacitor is zero. What happens when the switch is closed? The answer is indicated in the graph: the capacitor will charge at a rate that is fast initially but slows down as the charging proceeds. The shape of this graph is always the same, no matter what component values we use, but the amplitudes and time depend on the voltage of the supply and the values of resistance and capacitance that are used. The shape of the graph shows that the rate of charging decreases as the capacitor charges, so that, in theory, charging is never complete though for practical purposes the difference between the capacitor voltage and the supply voltage will be undetectable after some time. The simplest method of analysing what happens is by using the idea of time constant.

The time constant of a circuit like this is obtained by multiplying the capacitance value by the resistance value. If capacitance is measured in farads and resistance in ohms, the time constant will be in units of seconds, but it's more likely that you will work with times of milliseconds and microseconds, perhaps nanoseconds, as in Figure 9.19. The time constant corresponds to the time that the capacitor takes to be charged to a voltage level that is about 63 per cent of the final voltage, the supply voltage. This may appear to be a very awkward figure to choose, but it is not an arbitrary amount; it is the consequence of the shape of the graph for the charging capacitor. The significant point is that after another interval of time equal to one time constant, the

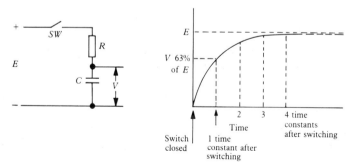

Figure 9.18 *Charging a capacitor through a resistor. For all practical purposes, charging is complete after a time equal to four times the time-constant*

Units of R	Units of C	Units of RC
Ω	F	seconds
k	µF	milliseconds (10^{-3}s)
M	µF	seconds
k	nF	microseconds (10^{-6}s)
k	pF	nanoseconds (10^{-9}s)

Figure 9.19 *The units of C and R for time-constant calculations. The use of R in k and C in nF is particularly useful for obtaining time constants in microseconds*

capacitor will have charged to about 63 per cent of the remaining voltage, which means to a level equal to about 77 per cent of the final voltage. After a time equal to four times the time constant, a capacitor will have charged to about 98 per cent of the final amount, so that for all practical purposes, we can say that the capacitor is fully charged.

When a charged capacitor is connected across a resistor, the capacitor will discharge, and the graph of voltage plotted against time for discharging looks as in Figure 9.20. Discharging follows the same pattern as charging, with about 63 per cent of the voltage discharged in a time equal to the time constant. As before, then, we can take it that a capacitor will be completely discharged in a time equal to four time constants. The significance of these charging and discharging time constants becomes apparent when we want to find out the effect of a capacitor-resistor circuit on a signal that consists of a sudden change of voltage.

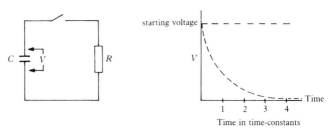

Figure 9.20 *Discharging a capacitor through a resistor. This follows the inverse of the charging curve, with about a 63 per cent reduction of voltage at each time constant step. Once again, discharging is virtually complete after four time constants*

Such a signal is called a **voltage step**, and signals of this type are used in all types of digital circuits, particularly in computer circuits. Figure 9.21 shows the shape of a voltage step from zero up to some fixed level, and the effect of the two possible simple capacitor-resistor circuits on this step. In each case, the waveform is the result of charging or discharging the capacitor, and the time that is needed is four time constants. In the first case, the capacitor is uncharged initially, so that each side is at the same voltage. After the step, however, the capacitor can charge, because one plate stays at the changed voltage and the other is free to pass current to earth through the resistor and so charge the capacitor. In the second case, the step voltage causes the capacitor to charge through the resistor until the voltage across the capacitor equals the voltage of the step.

These changes in waveshape may not necessarily be noticeable if the step up in voltage is followed by a step down, forming a square pulse. As Figure 9.22 shows, if the time between the step up and the step down is very short, much less than the time

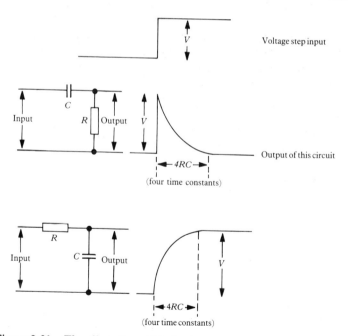

Figure 9.21 *The effect of two RC circuits on a voltage step*

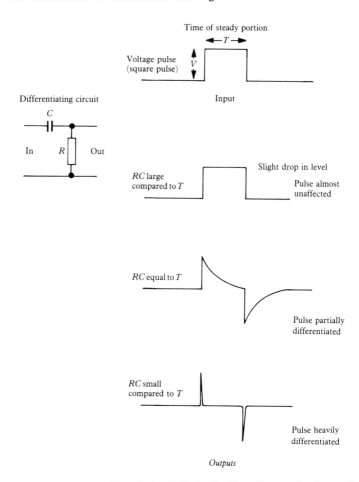

Figure 9.22 *The differentiating RC circuit. The effect on the shape of a pulse depends on the size of the time constant compared to the time of the flat top of the pulse. The overall effect is to emphasise the edges of the pulse*

constant of the capacitor and the resistor, then the shape of the square pulse is almost unchanged. When shorter time constants are used, however, the shape of the pulse is considerably changed, and with very short time constants, the square pulse is transformed into two pulses, one corresponding to each voltage step. This action of a time constant that is short in comparison to the time of a square pulse is called differentiation, and the resistor-capacitor circuit is a **differentiating circuit**.

AC circuits 133

The other form of circuit, as illustrated in Figure 9.23, has very little effect on the square pulse if the time constant is very long compared to the time of the pulse. Using shorter time constants means that the capacitor is given less time to charge, so that the voltage never reaches the upper step level, and so the pulse is distorted into a slight rise and fall of voltage. This form of circuit is called an **integrating circuit**, and its effect is to smooth out a pulse, reducing the voltage step and extending the time. Both of these basic circuits are extremely important in all types of circuits that involve pulses, from TV circuitry to computer operation.

Differentiation and integration can also be unwanted effects. Sending a pulse through a long cable will have an integrating effect, smoothing out the pulse and spreading out the time for which the voltage changes. Undesired capacitance between two

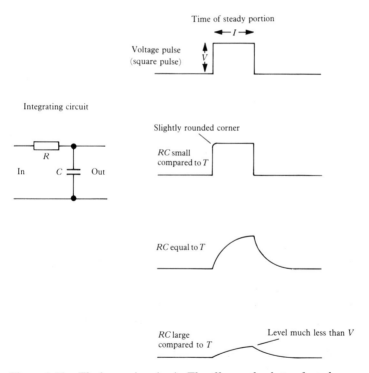

Figure 9.23 *The integrating circuit. The effect on the shape of a pulse once again depends on the size of the time constant compared to the time of the flat top of the pulse. The overall effect is to round off the pulse*

cables can result in a pulse on one cable producing a differentiated pulse on the other cable. In such examples, the capacitance is stray capacitance, and the resistance is the resistance that will exist in any circuit. One of the main problems in digital equipment is to maintain pulse shapes, particularly in a circuit that is physically large so that pulses have to travel along lines that are several centimetres in length. Many digital circuits make use of special circuits called **drivers** which are intended to reduce the problems of integration by supplying the pulse for a circuit that has very low resistance. This makes the time constant of this resistance with any stray capacitance very small, so as to have minimal effect on the pulse shape.

10
Electrons in a vacuum

The important feature that distinguishes electrons from holes as current carriers is that electrons can be set loose from a material and can exist and move in a vacuum. At one time, the whole of electronics was based on the movement of electrons in a vacuum, and the technology was that of the thermionic valve. Nowadays, valves form a very specialised technology of interest mainly in radio transmission. The release of electrons from hot materials, however, which is what the phrase thermionic emission means, is still used in the form of the cathode ray tube, and this forms a very important part of modern electronics. The different forms of cathode-ray tubes (CRTs) are fairly easy to understand if you know the fundamentals of electrostatics and electromagnetics, so that the main new feature is the release of electrons and their movement in a vacuum.

Cathode rays

To start with, a vacuum means a complete lack of any substance. In that sense, a real vacuum exists only in outer space, because the vacuum that we can produce on earth by means of pumps is a pretty poor substitute. The pressure of the air at the surface of the earth is due to the fact that the air covers the earth like a sea, and we live at the bottom of it. The pressure of this air will support a column of mercury some 760 mm high, so that millimetres of mercury are a convenient measure of pressure. By using vacuum pumps we can reduce the pressure inside a glass bulb to less than one hundredth of a millionth of a millimetre, and on the face of it, this looks like a good approximation to a vacuum. In terms of the number of molecules of gas that still exist in the bulb, however, it is nothing like a true vacuum. Despite this, the separation between the molecules will have increased to such an extent that an electron moving in the remains of the gas will strike a molecule, on average, only after travelling several centimetres. This average distance is called the **mean free path**. The pressure of the gas that remains inside a cathode-ray tube must be low enough to ensure that the distance that electrons will travel will be less than the mean free path for that pressure.

The simplest possible (and most useless) form of cathode-ray tube is illustrated in Figure 10.1. This consists of a glass tube, at one end of which is a fine metal wire, or **filament**, which can be heated by passing current through it. At the other end of the tube is a phosphor screen, a layer of a powder of a material which will emit light when it is struck by electrons. A ring of metal near this screen is connected to a terminal, the **anode**. A high voltage is connected between the anode and one end of the filament. When this voltage is connected so that the anode is positive, and when the filament is heated to a bright red, electrons are emitted from the filament. Since electrons are negatively charged, they will be attracted to the anode and in doing so will strike the screen, causing it to glow. There is no glow if the filament is not hot enough to emit electrons, or if the anode is not sufficiently positive to attract them to hit the screen at a high speed.

A cathode-ray tube for practical purposes needs to be able to

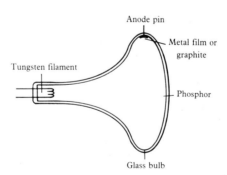

Figure 10.1 *A primitive cathode-ray tube. The phosphor coating on the screen will glow when the filament is white-hot and the anode is several hundred volts positive with respect to the filament terminals*

control the electrons by altering the amount and by changing their direction. The two main types of cathode-ray tubes are instrument tubes, used mainly for oscilloscopes, and TV or radar cathode-ray tubes, and the ways that they use to control the direction of the electron flow are different. In some ways, the instrument tube is easier to explain, so it makes sense to start with this type. To start with, the use of a hot wire filament to emit electrons is most unsatisfactory, because only a few materials, such as tungsten, can be used, and very high temperatures are needed. Fortunately, there is a mixture of metal oxides (of calcium, strontium and barium) which will emit electrons into a vacuum at much lower temperatures, at a dull-red temperature. This allows us to make **cathodes** that consist of a coating of these materials on one end of a metal cup. The metal cup cathode can be heated by an insulated wire, which can be of a metal like molybdenum, with the heater wire completely insulated and separate from the cathode cup. This allows us to make the shape and size of the cathode as we want it. We can also control the flow of electrons by surrounding this cathode with another metal cup, in which a small hole is drilled. (The arrangement is shown in cross-section in Figure 10.2.) If the voltage on this second metal cup, the **grid**, is sufficiently negative compared to the cathode voltage, then no electrons will move from the cathode. The electron flow is said then to be cut off. By reducing the amount of negative bias, we can control the electron flow, and the small hole on the grid will force the electrons to take paths that pass through this hole.

The path of the electrons in the rest of the tube can also be controlled by using voltages on metal cylinders (Figure 10.3). These voltages have the effect on an electron beam that a lens has

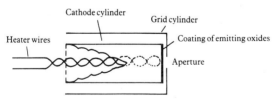

Figure 10.2 *The indirectly heated cathode of a cathode-ray tube. The oxide coating on the end of the cathode cylinder emits electrons when heated by the insulated wire spiral inside the tube. The intensity of the electron beam is controlled by the voltage on the grid cylinder*

138   Electronics for Electricians and Engineers

Figure 10.3 *A typical electron gun which will produce a focused beam of electrons at the face of a tube. Typical voltage levels are also shown*

on a light beam, so that the beam can be bent in such a way as to form a kind of image of the hole in the grid. When this image is situated on the screen surface, then the electron beam is focused and a spot of light will appear on the screen. The brightness of this spot can be controlled by altering the amount of bias voltage on the grid of the tube. The voltages that are applied to these electrodes are connected through wires that are sealed into the glass of the tube, because the tube must be evacuated for any electron beam to exist.

The direction of the electron beam can also be controlled. For instrument tubes this is done electrostatically, by making use of two sets of metal plates, (Figure 10.4). On the principle that a positive voltage on a plate will attract the electron beam and a negative plate voltage will repel the electron beam, the beam can

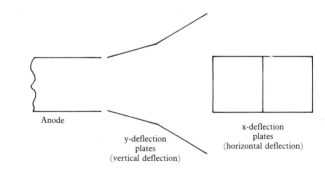

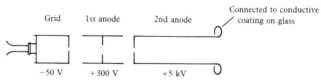

Figure 10.4 *The arrangement of deflection plates on an instrument CRT. The two sets of plates are set at right angles to each other, and the bent shape is to ensure that a highly deflected beam will not touch the edge of a plate.*

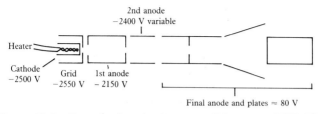

Figure 10.5 *Typical voltage levels on a small instrument CRT. The plates and final anode are at the DC voltage level of the signal output amplifiers. To provide a sufficient accelerating voltage, the cathode is held at a large negative voltage. An isolated winding is needed to supply the heaters with 6.3 V AC*

be deflected in two directions that are at right angles to each other. This allows the spot on the screen to be moved to any position on the surface of the screen, and the plates are called **deflection plates**. The tube will require a heater current for the wire that heats the cathode, a high voltage (the EHT, extra-high tension) voltage between anode and cathode, and suitable voltages for the focus and any other electrodes, and for the deflection plates. These voltages are higher than the voltages that are normally used in semiconductor circuits, and Figure 10.5 lists a typical set of voltages for a small instrument tube used in an oscilloscope.

The oscilloscope

The oscilloscope exists to display a picture of an AC waveform, and it is based on the use of an instrument cathode-ray tube. To display a wave, the beam is deflected vertically by one set of deflection plates, so that the spot will be oscillating vertically at the frequency of the signal that is to be viewed. Since electrons are for all practical purposes weightless, the beam can be moved with very little power required, even at very high speeds. At the same time as the beam is being oscillated vertically by the incoming waveform, it is also being moved horizontally by another waveform. This other waveform is a sawtooth (Figure 10.6), consisting of a steady rise of voltage (the **ramp**) followed by a fast return, the **flyback**. The effect of a sawtooth waveform applied to the horizontal deflection plates of a cathode-ray tube is to deflect the electron beam across the face of the tube at a steady speed and return it very rapidly. By convention, the beam is

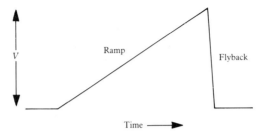

Figure 10.6 *A typical sawtooth waveform applied to the X deflection plates of an instrument cathode-ray tube*

always swept from left to right as you look at the face of the tube. The effect of this is to make the spot of the beam draw out a graph of the amplitude of the applied waveform as plotted against time. Unlike a graph drawn on paper, however, the graph that is drawn on the face of the oscilloscope tube is a dynamic graph showing the waveform as it exists at that moment. By altering the time of the ramp portion of the sweep waveform, the graph display can be expanded or contracted, allowing the waveform to be examined in detail, or compressed so that many waves can be displayed at one time.

The use of cathode-ray tubes in oscilloscopes demands the use of elecrostatic deflection using deflection plates. This is because the plates require no current other than the current needed to charge and discharge stray capacitances. This makes it easy to apply waveforms of widely differing amplitudes and frequencies, so that the main problem of oscilloscope design boils down to the provision of amplifiers for this range of frequencies. One problem, however, is the conflict between brightness and ease of deflection. To make a cathode-ray tube sensitive enough so that deflection can be carried out by amplifiers that work at reasonably low voltages, the accelerating voltage for the electron beam should be low. We need, however, to obtain a spot that is bright enough to see easily, and this demands a high accelerating voltage.

The conflict is solved by using the principle of **post-deflection acceleration** (PDA). One form of PDA (Figure 10.7) uses a spiral coating of graphite or high-resistance metal extending from the screen to a point near the deflection plates. The end of this coating at the screen is connected to a high voltage, and the end

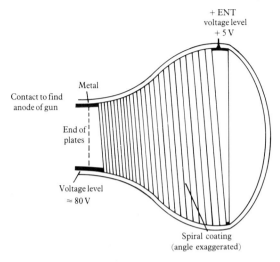

Figure 10.7 *One form of PDA. The electrons are accelerated after being deflected, so that the high accelerating voltage does not reduce the sensitivity of the deflection plates*

near the deflection plates is connected to a lower voltage. The deflection plates can then operate with high sensitivity, because the beam has been accelerated only by a comparatively low voltage up to that point. The trace on the screen can be bright because the beam has been greatly accelerated after being deflected. Another way of achieving this aim is to use a metal mesh separating the deflection plate side of the tube from the screen side.

Television CRTs

Cathode-ray tubes for TV use operate under totally different requirements. The rates at which the beam will be deflected across the screen and down the screen are fixed and will never be changed. The only change that is made is to the intensity of the beam, so as to alter the brightness of the spot on the face of the tube. The tube face will need to be large, usually considerably larger than that of any instrument tube, and the amount of electron beam current will have to be much larger than for an instrument tube, since most of the tube might have to display a bright portion of a picture. To see how these requirements are met, we need to look first at the simpler form of tube, a black-and-white (monochrome) tube.

To start with, the high beam current and high brightness are obtained by using a large accelerating voltage, of at least 10 kV, and up to 25 kV for colour tubes. At this sort of accelerating voltage, electrostatic deflection would require very large amplitudes of signals, so that TV tubes use electromagnetic deflection employing coils. These coils are placed outside the tube (Figure 10.8), so that their position can be adjusted, and the waveform through each set of coils will be a sawtooth *current* wave. This is not the same as a sawtooth voltage wave because each deflection coil is an inductor, and a step voltage waveform is needed to produce a sawtooth current wave in an inductor. The horizontal deflection rate is much higher than the vertical deflection rate, about 15 kHz for horizontal as compared to 50 Hz for vertical. The pulse voltage that is induced when current is cut off in the horizontal scanning circuit is so large that it is used to provide the EHT. But now we're getting into specialised details of TV circuitry.

Cathode-ray tubes for colour displays use the now-universal aperture grille principle. The screen of the tube is coated with successive stripes (Figure 10.9) of three different materials. These phosphors will glow with different colours, one green, one red,

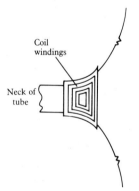

Figure 10.8 *The form of deflection coil structure used for TV tube deflection. The coil is basically a square section coil that has been bent around the neck of the tube*

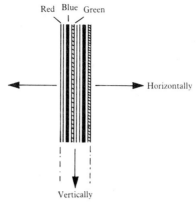

Figure 10.9 *The arrangement of phosphor stripes on the face of a colour CRT. The phosphors look identical when unlit, but will glow, respectively, red, blue and green when struck by electrons. The stripes are typically at 0.5 mm centres and are separated by very narrow non-fluorescent lines*

one blue. These three colours of light will together, and in the correct proportions, make up white light. By combining these light colours in other proportions, any natural colour, and quite a few unnatural ones, can be created. In place of a single electron gun giving a single beam, a colour tube uses three separate guns, arranged horizontally in line. One gun is arranged so that its beam hits only the stripes of phosphor that will glow red, the next is arranged so as to activate only the blue stripes, and the third will make only the green stripes glow. The guns are fed with different signals, depending on the amount of red, blue and green in the picture at each place, but all three beams are deflected together, and focused on to the screen very near to each other. The main problem then becomes to ensure that each beam strikes only the phosphor stripe that it is intended to strike. This is done by the aperture grille, a metal mesh that is used to shadow the phosphor stripes. As Figure 10.10 shows in principle, by placing an aperture in the metal grille close to the screen, the beams from the three guns can be forced to strike the correct phosphors. The metal grille can also be used as the final anode of the tube, and held at a voltage of around 20 kV. A lot of electron beam energy is wasted because a large part of the beam strikes the metal grille rather than the screen, and this is an inefficiency that we simply

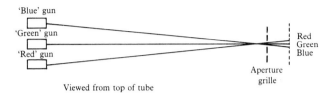

Figure 10.10 *The principle of using the aperture grille to separate the beams from three guns and so ensure that each beam hits only the correct colour of phosphor stripe in each set of three*

have to live with, because widening the apertures of the grille destroys the separation between the colours.

Though the deflection of a colour CRT follows along the same lines as that of a monochrome tube, additional windings and additional waveforms have to be used. For example, the three beams do not all stay in focus as the beams are deflected from the centre of the tube to one edge unless the focusing is changed as the beam is deflected. You can therefore expect a colour tube to make use of some of the scan waveforms in the focus circuits, and for this correction to be variable so that the tube can be set up for the best possible picture.

Radar requirements

For a lot of purposes, modern radar displays make use of TV techniques, with the pictures being processed by computers. The older and the simpler radar equipment often makes use of a rotating display, in which the beam is deflected from the centre of the tube to the edge, and at the same time, the position on the edge is gradually varied. The movement of the scan line around the tube is synchronised with the rotation of the radar aerial, so that the direction of a point from the centre of the tube corresponds to the direction of a radar target from the aerial. This type of display (Figure 10.11), is called PPI, **plan-position indicator**. At one time, the effect was achieved by using a deflection coil on the neck of the tube that was rotated along with the aerial. The alternative is to use coils that are fed with three sawtooth waves, and the phases of the waves are changed so as to rotate the beam as the aerial turns.

Electrons in a vacuum 145

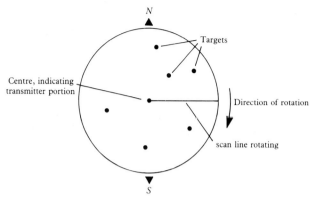

Figure 10.11 *A PPI (plan-position indicator) radar display. This consists of a scan from centre to edge of the screen, with the edge position rotated around the tube face in step with the rotation of the radar aerial. The distance of a target from the centre indicates its range, and the angle to the vertical indicates its bearing.*

11
The semiconductor diode

We saw in Chapter 1 that the materials that we call **semiconductors** can be engineered so as to have whatever value of conductivity, within limits, that we want. In addition, the conductivity can be mainly by electrons or mainly by holes, depending on what material is used to dope the pure (intrinsic) semiconductor. To avoid using long terms, the doped semiconductor that conducts mainly by electron movement (majority carrier electrons) is called **n-type**, and the opposite type (majority carrier holes) is called **p-type**. Neither, remember, has any permanent charge, and the names are used only to show which type of carrier takes most of the current through the material when current flows.

Junctions

Nothing is achieved by taking a piece of n-type semiconductor and placing it in contact with a piece of p-type material. The reason is that though electrons can pass from one to the other, holes cannot because a hole exists only within a single crystal. If we take a thin slice from a single crystal, however, and dope it from each side so that one side becomes p-type and the other n-type, then it is possible to make the types meet within the crystal, forming a **junction**. Such a semiconductor junction has properties that lie at the foundations of modern electronics, because it is rectifying – it will pass current in one direction only. The reason is not difficult to see if you know the fundamental rules of electrostatics.

Imagine a junction that has just been created. In Figure 11.1,

The semiconductor diode 147

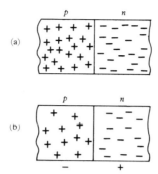

Figure 11.1 *A semiconductor junction imagined just as it is created (a) with holes as majority carriers on one side and electrons on the other. Immediately after formation (b) some carriers will have crossed over the junction and neutralised carriers of opposite charge. This leaves a new charge on each side of the junction that will oppose any further movement of carriers, and also leaves a 'depletion layer' with no carriers and hence acting as an insulator*

the + and − signs refer to the signs of the majority carriers, and do not mean that the material is itself charged. As the junction is formed, a few of these carriers will move across the junction, but because this movement leaves unbalanced charge on the side that the carriers have left, a small voltage will develop across the junction which will prevent any more carriers from moving. In the example, then, the *n*-type side of the junction becomes slightly positive and the *p*-type side slightly negative. The effect of this voltage difference will be to repel the charges away from the junction, so that the junction has no charges that can move. This very thin portion is called a **depletion layer**.

Now imagine what happens if a voltage is applied to the material from an external circuit, with a connection to each type of material. If the external voltage is connected so that the *p*-type material is to the positive side of the supply, the effect is to attract carriers over the depletion layer (Figure 11.2). This might not result in current flowing, however, unless the external voltage is greater than the voltage across the depletion layer that has been created when the junction was formed. Using external voltages up to the level of the internal voltage will make the depletion layer thinner, attracting charges towards the junction, and when the external voltage is raised to above the level of the internal voltage, current can flow. The electrons from the *n*-type material are

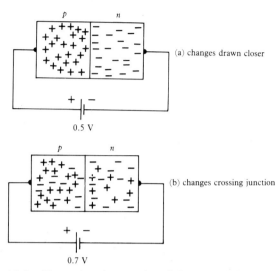

Figure 11.2 *How a junction transistor behaves with forward bias. This direction of voltage will draw the charges to the junction, making the depletion layer narrow (a). At a slightly higher voltage (b), the junction will conduct*

attracted into the *p*-type material by the positive supply voltage, and the holes from the *p*-type layer are attracted into the *n*-type layer by its negative supply voltage.

Now consider what happens if the external voltage is reversed (Figure 11.3). Instead of attracting the carriers into the depletion

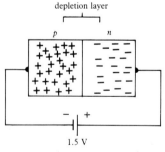

Figure 11.3 *Reverse bias will draw carriers apart, widening the depletion layer and preventing conduction. A very large reverse bias can cause minority carriers (electrons in P-type, holes in N-type) to accelerate and collide with atoms, splitting them into electrons and holes which will then cause conduction. This is the state of* breakdown

layer and then across the junction, the carriers in each type of material are attracted away from the junction. This makes the depletion layer larger, and ensures that the junction cannot conduct. The junction is now an insulator, and unless a very high voltage is used which will cause the minority carriers (holes in the *n*-type material and electrons in the *p*-type) to move, the device will behave as an open circuit. This simple arrangement of a semiconductor junction with a contact to each type of material is called a **semiconductor diode**.

Characteristics

A diode does not obey Ohm's Law, so that we cannot use $V = RI$ or any other simple formula to find a relationship between current and voltage. The relationship must be read from a graph, and a graph of this type is called a **characteristic**. A typical characteristic for a semiconductor diode is illustrated in Figure 11.4, assuming a diode made from *p*- and *n*-type silicon. The characteristic is really made up from two separate graphs. The right hand side is the graph for current flow, meaning that the

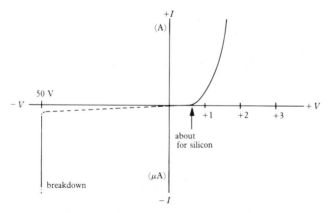

Figure 11.4 *A typical silicon diode characteristic. With forward bias, the diode starts to conduct at about 0.6 V, and the current increases steeply, not obeying Ohm's Law, as the voltage is slightly increased. With reverse bias almost unmeasurably small currents flow until the breakdown voltage is reached. Note the different scales of voltage and current used for the two sections of the characteristic*

diode has its *p*-type material connected to the positive side of a supply. This is referred to as the **forward direction**, and the external voltage is called a **forward bias**. The graph axes are scaled in terms of small voltages and currents, and the graph shows no current flowing until the forward voltage is about 0.6 V. This is the level of the internal voltage that was created along with the junction. When the diode starts to conduct, the graph shape is not the straight line that you would expect of a resistor, but a curve. The curvature is upwards, so that the current value is multiplied for each small increase in voltage. There is no constant value of resistance; the behaviour is as if the resistance decreased as the voltage was increased.

When the voltage is applied in the **reverse direction**, so as to make the junction non-conducting, the voltage scale on the graph has to be changed in order to show anything happening. For this particular diode, nothing is measurable until the voltage reaches a high value, around 100 V in this example, at which point the diode becomes conducting with a very low resistance. Unless there is some means of limiting the current flow, such as a resistor connected in series, the current that passes in this condition will destroy the junction. The reverse voltage that is needed to cause this effect is called the **breakdown voltage**. Applying breakdown voltage to a diode is harmful only if excessive current is allowed to flow, and the principle of reverse breakdown is used deliberately in Zener diodes, which we will consider later.

Another form of diode that is now extensively used is the **Schottky diode**. This has a junction that has been formed from a semiconductor in contact with a metal (usually aluminium). Unlike the conventional diode, the carriers in the Schottky diode (or Schottky barrier rectifier) are always majority carriers. This permits the diode to stop conducting very rapidly when it is reverse biased, an important feature in circuits that work with steep-sided pulses. Schottky diodes also have a smaller forward voltage across the terminals when they are conducting (as compared to conventional silicon junction diodes), and this makes them particularly suitable for high-current and/or low-voltage rectification. The principle of the Schottky diode is also important for some types of digital integrated circuits (see Chapter 14).

Rectification

One important use for semiconductor diodes is rectification, the conversion of AC into DC. We have already seen, in Chapter 7, how commutation can be achieved by using diodes, but commutation does not by itself achieve a DC output. The commutation of a single-phase AC supply, using four diodes in a bridge circuit, gives a unidirectional output that rises to a peak and falls again to zero at twice the frequency of the original AC. Using semiconductor diode bridges with three-phase AC, as in car alternator systems, gives an output that is closer to DC, because as one output voltage drops another will be rising to its peak. Even this scheme, however, produces too much 'ripple', i.e. a portion of the output still fluctuates in amplitude.

Rectification or commutation, then, has to be followed by a process which will reduce these fluctuations of amplitude. The simplest method is to use a capacitor connected on the output side of the diodes. A capacitor connected in this position is called a **reservoir capacitor**, and its task is to smooth the output by supplying the output current at a time when the diodes are not conducting. Each time the diodes of the rectifier conduct, there will be enough current at a suitable voltage to recharge the reservoir capacitor.

Figure 11.5 shows a typical circuit of diode bridge and reservoir capacitor, and also shows the voltage/time graph for the output. When the current taken by the load is negligibly small,

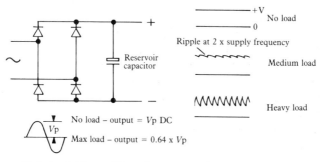

Figure 11.5 *A bridge rectifier and reservoir capacitor power supply, showing the effect of increasing load current on the supply voltage waveform. At maximum load, the DC output will be 64 per cent of AC peak voltage and the ripple will be of large amplitude*

the output voltage is almost perfectly smooth DC. When an appreciable amount of current is being drawn, however, the output contains fluctuations of lower amplitude. This is because the reservoir capacitor is supplying current during the time that the diodes are not conducting. When a capacitor supplies current, it does so by discharging, so that its voltage drops, and the greater the current the faster the voltage drop. The rate at which voltage drops, in fact, in volts per second, is equal to the current drawn (in amps) divided by the capacitance value (in farads). The voltage will rise again when the next diode conducts, but this alternate charging and discharging of the reservoir capacitor will cause a measurable ripple. This will normally be measured using an oscilloscope, and its value should be quoted as peak-to-peak, because the waveform is not a sine wave. The ripple is often, however, quoted as an RMS value, ignoring the difference in the waveform. The average voltage of output, as measured by a DC meter, is lower when the load current is high. This is because the presence of ripple reduces the average amplitude of the output. Figure 11.5 also shows the average values of DC for the extremes of load.

Many DC power supplies for electronics consist only of a transformer to obtain the correct AC value, a diode bridge for rectification, and a reservoir capacitor. For low-voltage supplies, very large values of reservoir capacitor can be used, typically 5000 µF or more. Supplies at higher voltages, 100–300 volts for example, will generally use lower values of capacitance, typically 16–100 µF. A low-voltage supply needs as large a reservoir capacitor as is feasible, because low-voltage supplies generally have to provide a large current, and any ripple is likely to be large compared to the output DC voltage. As it happens, large capacitance electrolytic capacitors can be made at low cost provided that the working voltage is low. The higher voltage supplies can use lower reservoir values because currents are usually low. For some purposes, such supplies may make use of paper or plastic capacitors which are very expensive in the larger capacitance values.

If the use of a reasonably large reservoir capacitor is not enough to smooth a supply to acceptable levels, then further smoothing, consisting of a large value inductor and another stage of reservoir, may be needed (Figure 11.6). If this also is insufficient, a voltage

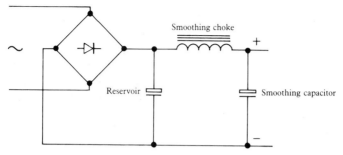

Figure 11.6 *Incorporating further smoothing with an inductor (usually termed a* choke *in this context) and a further capacitor*

regulator circuit will have to be used. This circuit requires an input at a voltage rather higher than the intended output, and its action is to control the output voltage to a steady level by regulating the current supply. For example, a voltage regulator or voltage stabiliser with an output of +5 V would require a supply with an output of +9 V. If the supply voltage dropped to +6.5 V, because of large load current, the regulator will still be able to supply +5 V, but this would not be possible if the supply voltage dropped to less than 6 V. Voltage regulators are dealt with in more detail in Chapter 13.

Multipliers

For a few purposes, a power supply may need to use a **multiplier** circuit. A multiplier is a method of obtaining a high voltage DC output from a comparatively low-voltage AC input, avoiding the need for a costly high-voltage transformer winding. The technique is particularly useful if a low-current high-voltage supply is needed in a circuit that otherwise operates at one single voltage. The circuit can, for example, be used in a computer power supply in which all but one part of the circuit runs with a +5 V supply, but one component needs a +12 V supply. The most common use of a multiplier, however, is in a colour TV receiver to obtain the +20 kV (approximate) supply that is needed for the screen voltage.

The principle is illustrated in Figure 11.7. The multiplier portion consists of diodes D_1 and D_2 and capacitor C_1. To

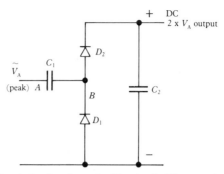

Figure 11.7 *A simple voltage-doubler circuit. The capacitor would normally be a paper or plastic dielectric type except for low-voltage low-current doublers which could use electrolytics*

analyse what happens, we need to think of the positive and the negative portions of the wave at point A from the transformer separately. During the negative half of the cycle, current will pass through D_1 and since there is no DC circuit present, will charge C_1. The sign of charge will be such that point B is positive, and the capacitor will charge to the peak value of the wave. D_1 does not conduct when the voltage rises again, and when the positive half cycle starts, point B will still be at the peak voltage, as is the output point C. The positive half cycle will make diode D_2 conduct, and the current will charge capacitor C_2. This capacitor will already have been charged to about the peak value of the wave because of the charge on C_1, so that this part of the wave adds the same voltage again, charging the capacitor to twice the peak value of the AC wave.

As always, taking current from this reservoir capacitor will reduce the voltage output and increase the ripple, but a multiplier like this is used only where the load current is low. The effect of load current on voltage can be clearly seen on some TV displays and many computer monitors in the form of a change of picture size when the brightness of the picture changes. This is a problem of poor regulation, and the regulation of a multiplier supply is very poor indeed. TV receivers often use a crude form of voltage regulator in order to stabilise the output from the multiplier stage that is used to provide the very high voltage supply (the EHT supply).

Demodulation

Demodulation means the recovery of a signal from a modulated carrier. The simplest form of demodulation is of an amplitude-modulated carrier, and the action is very similar to rectification using one diode. A typical AM demodulator circuit is illustrated in Figure 11.8, and the differences between this and the rectifier circuit are in values and time constants rather than in action. The diode will conduct only on the positive half-cycles of each carrier wave, and it will charge the capacitor C_1 to the peak amplitude of the carrier. The time constant C_1R_1 will be chosen so that the charge will leak to some extent, allowing the next positive carrier peak to charge the capacitor to a new value. The leakage must not be so great that the voltage falls almost to zero between carrier peaks, but not so small that the voltage cannot change rapidly enough. The time constant must be large compared to the time of a carrier wave but small compared to the time of the highest frequency of signal that will be modulated onto the carrier. This allows the voltage across the capacitor C_1 to follow the amplitude of the modulation rather than the amplitude of the carrier, and the components R_2, C_2 provide some further smoothing, removing traces of carrier frequency ripple. This type of demodulating circuit is far from perfect, but is judged sufficient for AM receivers.

The demodulation of FM is much more complicated. Most modern demodulators make use of a type of circuit called a **phase-locked loop**. This circuit produces a DC output which changes value when the frequency of the input signal changes

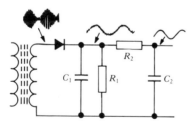

Figure 11.8 *A diode demodulator for AM signals. The time constant of C_1R_1 ensures that the voltage across the capacitor follows the outline of the modulation. The smoothing resistor and capacitor R_2C_2 remove remaining RF ripple from the output*

value. The output is positive when the frequency of the input signal is higher than a set frequency, and negative when the frequency of the input signal is lower than the set frequency. The action of the phase-locked loop depends on the use of diodes in the circuit, but the circuit does not have to be assembled – it is available in integrated form (see Chapter 13).

Clipping and clamping

The use of signals of pulse form creates a need for two more diode actions, clipping and clamping. **Clipping** is the removal of one part of a signal, usually a peak. Figure 11.9 shows a simple example of a clipping circuit that uses a diode, a voltage supply, and two resistors. The input is imagined to be a square pulse on top of which interference or capacitive pickup can add a spike of voltage. The diode is connected from the point where the resistors join, and is biased by a voltage roughly equal to the voltage of the square wave. During the time of a normal square wave, the diode will not conduct, because its anode is never at a voltage sufficiently greater than the cathode voltage. If a spike is present, however, the diode will conduct, and the voltage drop across R_1 will ensure that the spike voltage is reduced to about the normal voltage of the square wave, provided that the voltage V has been correctly set.

A **clamping** circuit is a much more elaborate device. Its purpose is to remove a low-frequency modulation that is affecting a pulse signal, usually because of interference from mains

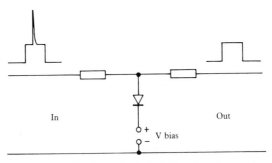

Figure 11.9 *Peak clipping. The diode is biased so that a voltage level above the normal maximum will cause the diode to conduct, shorting out the unwanted voltage peak*

circuits. A typical example is a set of pulses whose zero level rises and falls at 50 Hz because of hum interference from a mains transformer. A very simple type of clamp, called a **DC-restorer**, is illustrated in Figure 11.10. The principle is that the diode will conduct each time the voltage on the waveform is negative. This conduction will charge the capacitor, and the time constant of capacitor and resistor is long compared with the time between pulses, but short compared with the time of a 50 Hz hum cycle. The result is that the lower peak of each pulse is set to an almost constant level, so that the hum is removed from the signal. A true clamp circuit is more elaborate, and uses another pulse input to set the time at which the diode can pass current. Clamp circuits are widely used in TV circuits, particularly in TV camera circuits, to prevent the output from being affected by low-frequency interfering signals.

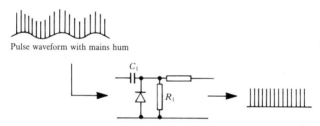

Figure 11.10 *DC restoration with a simple clamp. If the time constant C_1R_1 is suitably chosen, the capacitor will charge to a steady voltage, biasing the diode so as to remove the mains hum. A true clamp circuit uses transistors so that the conducting path can be opened or closed by applying a clamp-pulse at a specified time*

Zener diodes

A Zener diode is a diode having a reverse breakdown characteristic. The name is that of Clarence Zener, who discovered one of the reverse breakdown effects, but most of the diodes that bear the name of Zener do not, strictly speaking, make use of Zener effect but of avalanche effect, a more rapid form of breakdown. Whatever the actual effect, the Zener diode is used with reverse bias, and with a resistor in series in order to limit the amount of current that can flow. The voltage across the series combination of resistor and diode (Figure 11.11) is sufficient to cause

158 Electronics for Electricians and Engineers

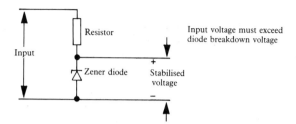

Figure 11.11 *The basic circuitry for using a Zener (voltage reference) diode. The diode is used reverse-biased so that the junction breaks down. The amount of current that can flow is controlled by the series resistor, and the voltage across the diode remains constant as the current changes*

breakdown, and since the breakdown voltage is almost constant regardless of the amount of current flowing (Figure 11.12), the voltage across the diode will be constant. By controlling the amount of doping in the manufacturing process, Zener diodes can be made in a 'preferred series' of breakdown voltages – the common values range from 2.7 V (2V7) to 75 V.

Diode A	Diode B
7.5 V, source resistance 5Ω	24 V at 0.5 A, source resistance 2Ω
Output is 7.5 V at 20 mA	Output is 24 V at 0.5 A
For each mA change in current, voltage changes by 5 mV	For each 0.1 A change in current, voltage changes by 0.2 V
Temperature coefficient 3 mV/°C	Temperature coefficient 18 mV/°C

Figure 11.12 *Characteristics of two Zener diodes compared. The slope resistance expresses the ratio of voltage to current after breakdown. For breakdown voltages below about 4 V, the temperature coefficient is negative, but for voltage levels above 4 V, the temperature coefficient is positive and increases for higher breakdown voltages.*

Varicap diodes

Varicap diodes make use of the change in the depletion layer of a diode as the bias is varied. When a diode is reverse-biased, the depletion layer widens as the reverse bias is increased, so making the separation between the conducting regions greater. The existence of the depletion layer means that the conducting regions are separated by an insulator, so that there will be capacitance between the terminals of the diode. Since the size of the insulating region changes with the amount of negative bias, the capacitance of the diode will also change, and a varicap diode is one whose doping has been controlled in such a way as to make the most of this feature.

A varicap diode is used as a variable capacitor, with the bonus that the variation is achieved electrically by changing the negative bias on the diode. They are used extensively in radio-frequency tuners, so that tuning can be accomplished electrically by using potentiometers rather than by the clumsy mechanical process of using variable capacitors. In addition, the use of varicaps makes automatic frequency control straightforward.

An automatic frequency control system uses a phase-sensitive demodulator whose output voltage will be zero at the correct frequency, but which will give a positive voltage if the input is at a higher frequency, or a negative voltage if the input is at a lower frequency. In either case, the voltage is proportional to frequency difference over a range of frequencies. The voltage from this form of demodulator can be added to a negative bias and applied to a varicap diode which is part of an oscillator circuit in a superhet receiver. This allows a signal to be tuned and then held, so that small variations in components of the oscillator circuit do not disturb the frequency. The use of automatic frequency correction of this type is essential for receivers working with the VHF or UHF range of frequencies, because oscillators for these frequencies are never sufficiently stable, particularly if tuning is critical as it is for colour TV reception.

Optoelectronics

Optoelectronics is the branch of electronics that deals with the conversion of signals between electrical form and light. Many of the optoelectronic devices that are in common use are, electrically speaking, diodes, and so are conveniently dealt with in this chapter. Some of the devices are very specialised, and are not mentioned here.

Starting with conversion of light to electrical signals, **photodiodes** are silicon diodes whose junctions are encased in transparent material. The diodes are operated with reverse bias, and the effect of radiation on them, including light radiation, is to supply enough energy to separate electrons from holes, and to make the depletion layer of the junction more conductive. This in turn will cause the (very small) reverse current to increase. Typically, a photodiode of this variety would have a **dark current** of about 1 nA with no illumination, and this could increase to as much as 1 mA under bright illumination.

The amount of current is proportional to the amount of light, but the colour of the light is also important, since the energy carried by a light wave is greater for short wavelengths (violet and ultra-violet) than for long wavelengths (red and infra-red). Transparent materials allow only a band of light frequencies to pass. The specification for a photodiode will include the peak spectral response, which is the wavelength of light for which the output will be a maximum for a given brightness. This wavelength is typically 750–850 nm. Silicon photodiodes are normally used along with amplifiers (such as a FET op-amp, see Figure 11.13) because of the small output from the diode, but their response time is fast, of the order of 0.5–250 ns, depending on construction, so that they can be used to convert rapidly changing light signals. This leads to applications in automatic counters for objects passing between the detector and a light source, readers for punched cards and tape, and receivers for light-signalling devices.

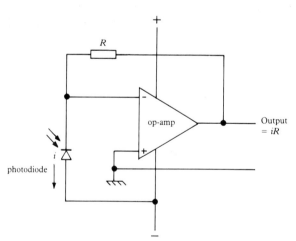

Figure 11.13 *A typical photodiode circuit using a FET op-amp (see Chapter 13) for amplification of the signals.*

Several other effects are also used. **Photovoltaic cells** give a voltage output that depends on light amplitude, and a true photovoltaic device, unlike a diode, needs no bias. The most common light detectors for use with steady or slowly-varying light, however, are photoresistors, otherwise known as **light-dependent resistors** (LDRs). These consist of tracks made from materials such as cadmium sulphide laid across an insulator, and covered by a transparent material. The resistance of the track depends on light amplitude, and can range from, typically, 10 M in darkness to around a few hundred ohms in bright light. The restriction is that the response of such detectors is very slow, measured in milliseconds rather than micro- or nano-seconds. For example, the ORP 12 is rated at 75 ms for resistance to rise after light has been cut off, and 350 ms for resistance to fall after light has been switched on; these are typical figures. For applications such as automatic light switches, fire detectors or flame detectors in oil-fired boilers, the speed of response is adequate and these devices are extensively used. Up to 100 V can be connected across the LDR, and a common type of circuit for operating a relay is illustrated in Figure 11.14.

The other aspect of optoelectronics is the conversion of electrical signals into light. The fundamental type of device for this purpose is the **light-emitting diode** (LED), which uses semiconductor materials that are not presently widely exploited for other forms of semiconductors. A typical material of this type

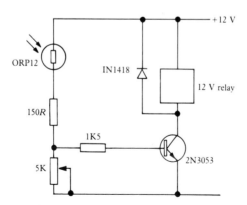

Figure 11.14 *A frequently used type of circuit for operating a relay from a photoconductive cell. Circuit courtesy of RS Components Ltd*

is gallium arsenide, a compound semiconductor (as distinct from a single element like silicon). All semiconductor diodes will radiate electromagnetic waves when current is passed in the forward direction. The radiation may not be in the visible range, or the packaging of the diode may prevent the radiation from escaping, so that a light-emitting diode is one that uses a light-emitting semiconductor and which has a transparent covering over the junction. The colour of the light that is emitted when current flows will depend on the type of semiconductor material, and the common colours are red and green (both together give yellow light).

The LED has a very high forward voltage compared to a silicon diode, of the order of 2 V, so that the supply must be adequate – you cannot expect to operate LEDs from a 1.5 V dry cell, for example. The brightness of the emitted light depends on the amount of current that is passed, and a figure of about 20 mA for a small emitter is typical. Single LEDs are used extensively as panel indicators, and it is possible to buy tri-colour LEDs which will give red, green or yellow light according to the terminals that are activated. Another combination device allows the colour to be changed by reversing the polarity of the supply. When a single LED is used, however, precautions should be taken to ensure that the polarity of the supply is never reversed, because the reverse breakdown voltage of LEDs is very low.

LEDs are packaged into displays of various types, such as bar and digit displays. The bar arrays use LEDs in the form of short bars laid either horizontally or vertically, and the principle is to indicate some quantity such as signal amplitude, fuel tank contents, temperature, etc. by means of the length of the illuminated section. Bar numbers of 10–30 are commonly used, and the LED displays can be bought combined with a display driver that will carry out the switching of the bars in response to the amplitude of an input signal. The other common use of LEDs is in seven-segment displays, (Figure 11.15). As the name suggests, these use seven bar-shaped LED segments that can be illuminated selectively in order to display the numbers 0–9, and a few letters. Many of these displays contain an eighth segment, a decimal point which can be placed either on the left or the right according to the type number of the display. These seven-segment displays are available in a range of sizes from 0.3 in to

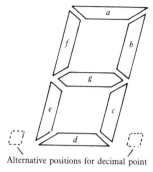

Figure 11.15 *The seven-segment display. Despite the name, many are eight-segment displays that include a decimal point*

1 in, depending on the distance at which they will have to be read. The displays are normally driven by ICs so that the decoding of the segments is carried out inside the IC, and the input to these ICs is normally in the form of a four-bit binary number. Combinations of LED and counter can also be obtained, and sets of seven-segment displays can be bought along with all the ICs that are needed for a counter of multi-figure display all in one package.

The disadvantage of using LEDs is that the power consumption can be fairly large, and this can be prohibitive for battery-operated equipment. Displays that consist of miniature filaments can also be used, and can offer a brighter display with lower power consumption. For battery operated equipment, however, the universal display type is the **liquid-crystal display** (LCD). Each LCD segment consists of a cell that is electrically a capacitor rather than a diode, so that the current required is negligible. The display does not, however, emit any light, and its visibility depends on some source of light being present. For this reason, if the LCD is to be used in dark conditions, some form of lighting must be supplied, and some displays feature back-lighting using miniature filament bulbs.

The LCD consists of a reflective conducting plate on which a thin film of the liquid-crystal material is applied. The front cover consists of a transparent panel whose inside surface is a conductor. A light-polarising filter is usually incorporated in this front panel, or laid on separately. The liquid-crystal material consists of molecules that are in the shape of long strings, and these polarise light, passing only the light whose electrical wave vibration is along the length of the molecules. This is also the action of the polarising filter, so that if the two are polarising light in the same direction, the light passes into the cell and is reflected out again. Applying an electric field to the cell, however, will turn the molecules round so that the light polarisation changes by 90 degrees. This prevents any reflection, so that the cell appears dark. The display therefore consists of dark segments on a light background.

The cell operates correctly only when the supply is AC, and it is unusual to supply cells individually. Displays are usually packaged in groups of digits or alpahabetical characters, and with all the AC power supplies and driving logic integrated, so that the unit can be used directly from a DC supply (typically 5 V). The current required for a complete display will normally be only a few microamps, as compared to several milliamps per segment for LEDs.

Opto-isolators

Opto-isolators (Figure 11.16) are combinations of LEDs and photodiodes in a package that allows light to pass from LED to photodiode, but which keeps the devices electrically isolated. The

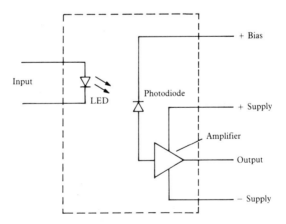

Figure 11.16 *The opto-isolator. There is no electrical connection between input and output, only a light path from the LED to the photodiode*

device is intended to allow comparatively small signals to be passed between circuits that are at very different DC levels. The advantages compared to the use of a transformer are that DC signals can be passed, and the bandwidth can be very wide. One typical application is passing cathode modulation signals to an instrument cathode-ray tube which might be at a voltage of − 3 kV. Opto-isolators are extensively applied in circuits that use thyristors (see below) to control mains voltage, because the use of an opto-isolator allows the controls for the thyristor circuit to be at low voltage levels. Note, however, that some electrical wiring regulations do not permit the use of opto-isolators for such purposes.

Triggered diodes

The simple semiconductor diode will pass current when it is sufficiently forward biased, and will not pass current when the bias voltage is reversed. Diodes can be manufactured with more than one junction, and these types of **multilayer diodes** allow the action of the simple diode to be modified. The most useful is the **thyristor**. This consists of three junctions arranged as pictured in Figure 11.17(a), and with a connection to the intermediate *p*-type layer as well as to the normal anode and cathode. The connection to the intermediate layer is called the **gate**, and its purpose is to

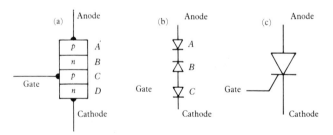

Figure 11.17 *The thyristor junction arrangement (a) and the equivalent diode arrangement (b) and thyristor symbol (c). Making the junction between gate and cathode conduct, switches charges into the other pn junction; with no connection to this junction, the charge remains and keeps the junction conducting for as long as the voltages are of the correct polarity between anode and cathode and there is sufficient current to maintain a flow of charge. The pulse at the gate need be of only a few μs duration*

make the device conduct between anode and cathode. In crude terms, you could think of the thyristor as being three diodes connected as shown in Figure 11.17(b). On the face of it, you would not expect this to conduct because of the central diode which is connected in the opposite sense, reverse biased, when the main anode is positive and the main cathode negative.

The action is rather more complicated than this simple equivalent shows, because the two central layers are very thin and close to each other. When the gate connection is made positive, even if only briefly, current will flow between the gate and the cathode, so that electrons move from layer D into layer C. Because the layers are so thin, however, many of these electrons will travel straight on into layer B, during which time the junction between layers B and C will be conducting. Now because of the voltage on the main anode, this will allow current to keep passing across this junction, so that the depletion layer cannot be restored. The device now conducts like an ordinary diode. It will continue to conduct until the voltage between anode and cathode is reduced to a level that does not allow conduction (or reversed), or until the current is reduced to a very low level, insufficient to keep the depletion layer from forming. When either event happens, the thyristor becomes non-conducting once again.

The thyristor, then, is a controlled diode which will conduct only when the anode is sufficiently positive with respect to the cathode (about 1 V) *and* a brief positive pulse is applied to the gate electrode. The action is used for any device in which a trigger action is needed, such as firing flash tubes for photographic use, but also for controlling the speed of AC motors. To see how this can be done, consider the waveforms of Figure 11.18. In this example, an AC wave is shown with a trigger pulse that is timed to occur at the peak of each positive half-cycle. If the AC wave is applied to the anode of a thyristor, and the pulse to the gate, then the output at the cathode will be as shown, a sharp rise to peak voltage at the time of the trigger pulse followed by a fall to zero voltage following the shape of the AC wave. By varying the timing of the trigger, we can control how much of the AC half-wave that passes, and if a pair of thyristors is used then both halves of the cycle can be controlled. This allows machines such as small motors and lamps to be controlled with none of the waste of power that would occur if a series resistor were used. A **triac** is a

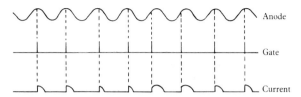

Figure 11.18 Power control with a thyristor. By altering the phase of the pulse at the gate, the thyristor can be turned on at different times in the AC cycle, allowing the average current through the thyristor and load to be varied

device that consists of two thyristors connected back-to-back so that the whole AC wave can be controlled.

Another form of four-layer diode is the **silicon-controlled switch** (SCS). This follows the same pattern of four layers, but with a separate connection to each layer. The SCS can be used as a thyristor by making use of the anode, cathode and one intermediate layer connection, or in other ways (as a unijunction) by making use of the other intermediate layer connection. The device is used in only a few specialised applications and will not be dealt with in detail here.

12
Transistors

Transistors are at the heart of all modern electronics, and in this chapter we will look at the more important types, their qualities and their applications.

The bipolar transistor

The **bipolar transistor** is a device that makes use of two junctions in a crystal with a very thin layer between the junctions. This thin layer is called the **base**, and the basic type of transistor depends on whether this base layer is made from p-type or n-type material. If it is of n-type material, the transistor is said to be p-n-p type, and if the base layer is of p-type material, the transistor is said to be an n-p-n type. The differences lie in the polarity of power supplies and signals rather than in the way that the transistors act. For most of this chapter, we shall concentrate on the n-p-n type of transistor, simply because it is more widely used.

Imagine a sandwich of layers as illustrated in Figure 12.1, with the middle (base) layer of p-type material. Of the other two layers, one is usually of larger area and is called the **collector**, and the other is the **emitter**. In a working circuit, a steady voltage would be connected between the collector and emitter, with the collector positive. With no connection made to the base, no current will flow between the collector and the emitter because the junction between the collector and the base is reverse-biased.

If the base is biased, all this changes. When the positive voltage on the base is sufficient to allow current to pass between the base and the emitter, the current will consist of electrons moving from

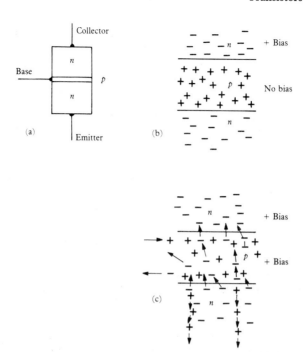

Figure 12.1 *The construction of an n-p-n transistor (a). The voltage between collector and base normally keeps this junction non-conducting (b), but when current flows in the base-emitter junction (c), most of the moving charge will cross the collector junction also. The collector current is very precisely proportional to the base current over a very wide range. The p-n-p transistor is similar except that the order of layers and voltages is reversed*

the emitter (emitted, hence the name) into the base layer. Because the base layer is so thin, however, most of these electrons will move on across the collector-base junction, making this junction conduct. Unlike the thyristor action, however, this current is always proportional to the base current; a typical value is that the current between emitter and collector is some 300 times the current between base and emitter. The exact value of this ratio depends on how thin the base layer is, and it varies considerably from one transistor to another, even when the transistors are mass-produced in the same batch. This value is variously known as **forward current gain** or **Hfe**.

The action of the *p-n-p* transistor is similar, except that the holes are the charge carriers, and the polarities are changed, with the collector negative, and a negative bias voltage applied to the base. Whichever type of bipolar transistor is used, it will involve the movement of minority carriers through the base of the transistor, and this is the factor that in many ways can limit the usefulness of a bipolar transistor for fast switching or for power output. One drawback is **carrier storage**. When the base bias is abruptly turned off, the base region still contains minority carriers and so the transistor will still be conductive until these carriers are removed. A sudden reversal of base bias in a transistor may therefore allow excessive reverse base current to flow for a short period.

The other problem is **secondary breakdown**. In the base region of a bipolar transistor, where the current between emitter and collector is carried by minority carriers, the temperature coefficient of resistance is negative. If any current path becomes overheated, therefore, it will also become more conductive, take more of the current, and overheat further until destruction. This is sometimes called 'current-hogging', and it also makes it impossible to share current evenly between two bipolar transistors connected in parallel. The secondary breakdown of bipolar transistors confines the operation of the transistor to a safe operating area which means that the actual power that the transistor can handle is usually much lower than would be suggested by the figures for maximum voltage, maximum current and maximum dissipation.

The FET

The **field-effect transistor**, or FET, is as old a type of device as the bipolar transistor but was commercially exploited much later. The principles are very different, and they involve a current path which consists always of majority carriers, unlike the bipolar transistor in which carriers must always move through a region (the base) in which they are in a minority. In addition, the input to a FET is either to a reverse-biased diode or to a terminal which is almost an open-circuit. This means that the input resistance is very high, and is reasonably constant, unlike the 'conducting diode' input of the bipolar transistor. Once again, though, the operating principles of the FET depend mainly on the simple rules of electrostatics.

University of Strathclyde Libraries
Main Library
CheckOut Receipt

04/10/04
03:12 pm

Item:Electronics for electricians and
engineers / Ian R. Sinclair.
Due date (yyyy-mm-dd) :

2004-11-15 21:00:00

Thank You for using

Please retain this receipt

The predominant FET type is the **MOSFET**, so that this will be described in more detail. An FET allows current to flow between two terminals, the **source** and the **drain**, which are connected by a thin strip of doped semiconductor called the **channel** (Figure 12.2). All of the semiconductor material between the source and the drain terminals will be of the same doping, which can be either *p*-type or *n*-type, though *n*-type is more common for the channel. The choice of doping determines the type of carrier that will be used, so by selecting *n*-type doping, the carriers will be electrons. The operating principle of the FET is to control a depletion layer in the channel so that the channel's ability to conduct can be altered. If, for example, a depletion layer

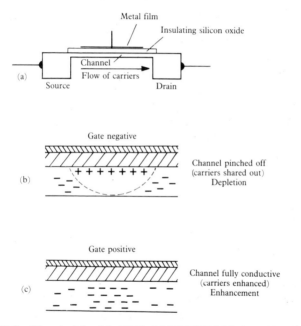

Figure 12.2 *The principle of the FET. A MOSFET (a) is formed by placing a layer of insulation, then a layer of metal, over a thin narrow channel of semiconductor, usually n-type. A voltage on the metal will induce charge of opposite sign on the channel and this will either trap carriers (b) or inject extra carriers (c). The current between source and drain is therefore controlled by the voltage at the gate, but no DC current flows in the gate since this is one terminal of a capacitor*

exists all the way across the channel, conductivity would not be possible and no current would flow. This is described as being **pinched off**, and the degree of pinching will control the conductivity of the channel by controlling the amount of depletion. The opposite effect, called **enhancement**, is also possible, using an electric field to inject more carriers into the channel so that it conducts more easily.

The differences between the main FET types are concerned with how the channel is enhanced or depleted. In general, a FET will be constructed to be used either in enhancement or in depletion mode. If it is designed to use enhancement mode, then with no bias voltage between the third (gate) terminal and the source, the FET will pass no current. A bias *voltage* (not current) must be applied in order to make current flow in the channel. For a FET operated in depletion mode, the current with no voltage between the gate and the source is the maximum, and bias must be applied in order to reduce this current. In the MOSFET, the bias voltages are applied to a metal contact that is deposited over a thin insulating layer over the channel. The materials used are, in order of gate to channel, metal, silicon oxide, and the semiconductor of the channel, so providing the MOS of the title. Enhancement mode is more common, because this allows devices to be operated in a 'fail-safe' way: if the bias fails, then the devices do not pass current.

Silicon oxide is an excellent insulator, so that the resistance between the metal gate terminal and the semiconductor channel is very high, hundreds or thousands of megohms. For practical purposes, then, we can regard the gate terminal as having infinite resistance, an open circuit, so that a DC bias voltage applied to this teminal will cause no gate DC current to flow. As far as signals are concerned, however, the gate is one terminal of a capacitor, and there will be a measurable capacitance of around 4–10 pf between the gate and the channel. This capacitance must be taken into account when the signals are at such high frequencies that this size of capacitance represents a significantly low reactance. For frequencies in the audio and low radio frequency range, however, we can think of the gate terminal as being an open circuit.

The voltage on the gate will affect the current that passes through the channel. The voltage that is applied between the drain and the source has very little effect on the channel current. The channel current is controlled almost completely by the voltage on the gate, so that provided the drain-to-source voltage is not allowed to become too low or too high, only this effect need be considered. The usual way of measuring the effect is to quote the figure of **mutual conductance**, written as 'Yfs' or, more correctly, 'gfs'. The difference is that g is the inverse of impedance Z whereas Y is the inverse of resistance R. At low frequencies, the difference is negligible. Either figure is usually quoted in micro-siemens (μS) or milli-siemens (mS), which means current divided by voltage. One mS is one milliamp of current change in the channel per volt change between the gate and the source. If, for example, the FET is quoted as having gfs of 1500 μS, this means 1.5 mS, or 1.5 mA per volt. This figure is used to find the voltage gain, because the figure of gfs in mS multiplied by the load resistor value in K will give voltage gain. For example, if we have a gfs value of 1.5 mS and a load resistance of 10 K, then the voltage gain will be $10 \times 1.5 = 15$ times. This is a modest figure of gain as compared to a bipolar transistor, and it is one reason for the continued use of bipolar transistors.

In circuits that make use of separate components (discrete circuits) as compared to ICs, the use of FETs is confined to specialised actions. These include amplifiers with high input resistance for measuring instruments, and mixers for very high radio frequencies. The mixer application arises because of the ease with which FETs can be made with more than one gate. A dual-gate FET will permit the channel current to be controlled independently by the two gates, and by applying an incoming RF signal at one gate and an oscillator signal at the other, the signal from the drain will contain the frequencies that result from mixing. More significantly, however, there is virtually no connection between the oscillator circuit and the RF circuit, so that interaction is minimised. Mixer stages using bipolar transistors at high frequencies suffer from oscillator 'pulling', in which the frequency of the oscillator is affected by the incoming RF signal because the isolation between the circuits is poor. In addition, there is a risk of the oscillator frequency being radiated out from the receiving aerial. The use of a dual-gate FET almost eliminates these problems.

Power MOSFETs

In addition to the specialised use of MOSFETs for small signals, MOSFETs of specialised construction are used for power output. Power output from an electronic circuit implies that a signal voltage must be maintained across a load that has a comparatively low resistance, and possibly a low reactance as well. This requires devices that can pass large currents, usually at fairly high voltage levels. Typically, a power output device might need to be able to work with 100 V between drain and source, and with currents that could be as high as 12 A, though the maximum current would not pass when the full rated voltage was applied, because the dissipation limit would not allow this. Dissipations of 20–125 W are typical for such power MOSFETs, and the figures for gfs are large, measured in siemens rather than in milli-siemens. A figure of 1–6 S is typical for such devices. A gfs figure of 1 S means that for each volt change of signal amplitude at the gate, there will be 1 A change of current between source and drain. The very large figures for gfs do not imply that the voltage gain in a circuit will be very large, because these devices are intended to be operated with low load resistance values, in the region of a few ohms only. The voltage gain will be equal to the gfs figure in units of S multiplied by the load resistance in units of ohms, so that for a 4 ohm load and 1.5 S, the voltage gain would be $4 \times 1.5 = 6$ times. The power gain, however, is enormous, because the power dissipated in the load will be many watts, but the power dissipated at the gate is negligibly small because of the very low gate signal current. This contrasts with the bipolar type of transistor which requires quite large base currents for power transistors, with figures of 1 A or more often needed.

Power MOSFETs are used extensively for driving signal into low resistance devices, for fast switching of large currents, and for high frequency amplifiers that must pass large currents. The advantage of FETs for such purposes is that there is no secondary breakdown in FETs; current is shared rather than hogged because the temperature coefficient of resistance in the channel is positive (because majority carriers are flowing). This means that the safe operating area for power FETs is much larger than for bipolar transistors, and the main limitation is the permitted dissipation. Figure 12.3 shows some typical power FET circuit principles.

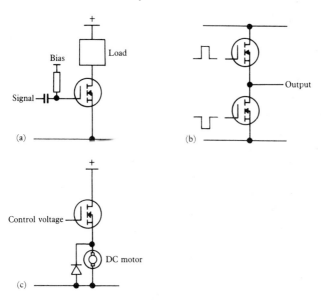

Figure 12.3 *Typical uses of power MOSFETs as power audio amplifiers (a), switches for fast-changing waveforms (b) and as motor controllers (c)*

VMOS devices

VMOS means **vertical MOS**, and the name is applied to a type of power MOS construction in which the gate is a V-shaped groove in a crystal, so that the channel runs down the groove (Figure 12.4) rather than along the surface. The main advantage of this type of construction is that the speed of operation can be very much faster than that of a conventional power MOSFET, though the ratings of voltage, current and power are all lower. A typical figure is a switch on or switch off time of about 5 ns (0.005 μs),

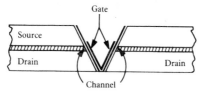

Figure 12.4 *The VFET principle. The drain can be in contact with the metal case of the FET so as to ensure good thermal conductivity*

which contrasts with times in the 50–300 ns region for conventional power MOSFETs. VMOS devices are therefore applied to ultra-fast switching and to amplification at very high frequencies.

JFETs

Junction FETs, or JFETs, use the same principle of a channel between drain and source whose current is controlled by voltage at a gate terminal. The gate terminal, however, is one end of a diode formed over the channel. If the channel is made from n-type material, as is usual, then there is a p-n junction formed near the source end of the channel, and the connection to the p-type material is the gate terminal of the JFET. For most applications this junction is reverse biased, so that the resistance between gate and source is high. For a lot of purposes, then, the JFET behaves rather similarly to the MOSFET, with negligible current flowing between gate and source. One important difference is that the gate is not so isolated, and damage due to electrostatic voltages is much less likely. The other difference is that if the bias voltage changes to a level that allows the junction to be forward biased, then appreciable current will flow between the gate and the source.

Switching with FETs

FETs have specialised applications in linear circuits such as amplifiers, but their main applications are in switching (digital) circuits. The reasons are not difficult to appreciate, because the FET is an almost ideal switch. With the gate biased off, the current that flows between source and drain can be quite negligible, a matter of a microamp or so. With the gate biased on, the resistance between source and drain can be low, around 100 ohms for a small FET or as low as 0.2 ohms for a power type. The main advantage, however, is that no current flows at the gate terminal other than the charge and discharge current of the small capacitance between the gate and the source. Figure 12.5 shows two typical switching circuits.

Circuits

All of the circuits in which transistors of either type are used can be

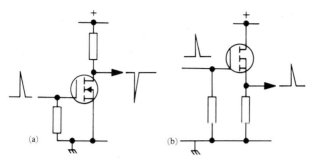

Figure 12.5 *Basic switching circuits, assuming* n-*channel enhancement MOSFETs: (a) pulse inverter, (b) source follower. Note that any MOSFET circuit must have a resistive connection between gate and source to avoid damage from electrostatic voltages*

understood if the basic facts of current gain (for bipolar) and gate action (for FETs) are understood, because these are the fundamental actions of these transistor types. Unfortunately, a large number of textbooks have survived from the very early days of transistors and they make what is a comparatively simple action appear very hard to understand. The remainder of this chapter, then, is concerned with how the simple facts can be harnessed to achieve all the actions that are now possible with semiconductor devices.

Loads

In a few applications, the current control of the bipolar transistor or the current switching of the FET can be used directly. A device that needs a comparatively high current, such as a relay for example, can be operated with a very small current by making use of a transistor in a circuit such as that of Figure 12.6. The relay is connected between the supply positive and the collector of the transistor. The diode is used to short out the large negative pulse that will be caused when the relay is switched off, because at switch-off, the change of current through the relay coil will induce a voltage which could destroy the transistor. The FET is less sensitive to this reverse voltage, but should nevertheless be protected.

The relay will not operate unless the base of the bipolar transistor passes current, enough to make the collector pass the

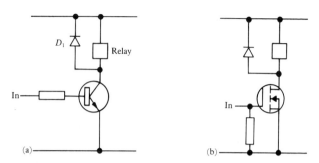

Figure 12.6 *Switching a relay with a bipolar transistor (a). The diode is used to protect the transistor from the negative voltage induced in the relay coil when current is switched off. The FET equivalent (b) is similar in structure*

current that is needed to operate the relay. If the relay needs 100 mA to operate, however, and the transistor has a current gain of 300, then the current that is needed at the base will be 1/3 mA, making the action of the relay much more sensitive than it would otherwise be. The voltage between the base and the emitter will be around 0.6 V. This is never exactly predictable, and is usually in the 0.6–1 V range for a wide range of base current values. The usual way of ensuring correct action is to use a supply for the base that will give at least 1 V, and a series resistor to limit the current to a safe amount. For the FET circuit, only a small voltage change will be needed at the gate to turn the relay on or off.

Voltage amplification

A transistor that has any form of load connected to the collector will permit voltage gain or amplification provided that the base of a bipolar transistor is passing current. To see why, consider the circuit of Figure 12.7. In this circuit, the resistor R_2 has a value of a few K and R_1 has been chosen to pass a small current, enough to set the collector current at a few mA and so make the voltage at the collector equal to 6 V, half of the supply voltage. Now consider the effect of a small voltage signal at the base. In the positive half of the input signal, the base current will be increased by the rising voltage, and so the collector current will also be increased – but by, say, 300 times as much. This will

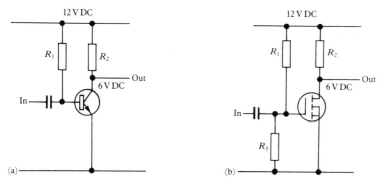

Figure 12.7 *A basic voltage amplifier stage. The bipolar version (a) is not very practical because the value of* R_1 *is very critical. The FET version (b) is a more practical arrangement. The amplifier inverts a signal and provides gain which is very much greater when the bipolar transistor is used*

make the collector voltage *drop*, because increasing the current through R_2 will make the voltage across R_2 greater, and so make the collector voltage lower. On the negative half cycle of the input wave, the base current will be reduced, the collector current reduced and the voltage at the collector increased. If the current in the base circuit were perfectly proportional to the voltage, then the result would be a wave at the collector that was a perfect copy of the wave at the base, but of many times the amplitude (40–100 times, typically).

This is a situation of voltage gain with inversion. The gain refers to the increase in signal amplitude, and the figure of gain is equal to output amplitude/input amplitude, with both measured in the same way (usually peak-to-peak). The output signal is inverted because an increase of voltage at the base causes a decrease of voltage at the collector. The FET equivalent of this circuit uses a simple voltage divider to bias the gate (enhancement mode assumed here), and the voltage of the input signal will control the current between the source and the drain. The gain will be lower, but in all other respects the action is similar, so that the signal will be inverted by this circuit also. The main difference between this and the bipolar circuit is that the gate of the FET takes no current.

An arrangement such as has been illustrated in Figure 12.7(a) is a single-stage common-emitter transistor amplifier; Figure 12.7(b) shows a single-stage common-source FET amplifier.

Single stage means that only one device is carrying out the amplification of signal, and **common-emitter** means that the emitter terminal of the transistor is common to both the input circuit (since it passes base current) and output (since it passes collector current). The arrangement of Figure 12.7(a) is not very practical because of the difficulty of ensuring that resistor R_1 has a value that will supply exactly the correct amount of base current, or **bias**.

A more useful arrangement is illustrated in Figure 12.8. This uses a potential divider to provide a voltage of, typically, 1–1.5 V at the base. The amount of current that flows between collector and emitter is then controlled by resistor R_4, which will be a comparatively low value in the range 100R–1K5. The merits of this system are that you don't have to make any assumptions about the current gain figure of the transistor, and changes in current gain have very little effect on the bias. The value of R_4 is set as shown in Figure 12.8, and the only assumptions that are made about the transistor are that its base current is small, and the base-emitter voltage will be about 0.6 V. Both will be true for all but a few silicon transistors. When a bipolar transistor is operated with a load R_3 (unit: K) and with a steady collector bias current of I mA flowing, then the amount of voltage gain is approximately $40 \times I \times R_3$. This figure is the (signal voltage out)/(signal voltage in) for small signals, i.e. signals that do not cause the collector voltage to reach the extremes of supply voltage or earth voltage.

One important point is the capacitor across R_4. If this is not included, then when a signal is applied to the base there will also be a signal at the emitter. This signal will be of almost the same amplitude as the signal at the base, and in the same phase. The effect will be to reduce very greatly the amount of signal that is applied between the base and the emitter, so that the gain of the transistor appears to be very much less. This is a form of **negative feedback**, and its effect is to reduce gain and reduce distortion. The gain of this type of circuit with the capacitor omitted is about the same as the ratio of resistors R_3/R_4. The output wave will, however, be a better copy of the input (lower distortion), and the circuit can cope with a wider range of input signals without overloading. The higher gain that can be obtained by using bipolar transistors makes these the preferred type for voltage amplifiers.

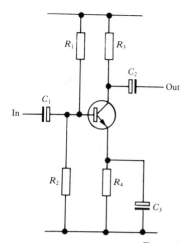

Calculations

1. Set base voltage by suitable values of R_1, R_2. Aim to pass about 1 mA through R_1, R_2 for a voltage amplifier. Set base voltage to about 1 V – 1.5 V

2. Calculate value of R_3 to set collector voltage to half of supply voltage, assuming a collected current of a few mA. Often better to pick a value for R_3 of 2K2–5K6 and then calculate collected current

3. For this value of current through R_4, emitter voltage must be about 0.6 V less than base voltage. Calculate R_4 to suit.

4. Make C_3 so that reactance of C_3 at lowest frequency to be used is less than value of R_4

5. Max. gain is about $40 \times 1.27 \times 4.7 \approx 238$, but this will be reduced by the potential divider action of R_1, R_2 on the signal through C_1, and by potential divider action of the load at the output

Example
12 V supply
$R_1 = 1K5$
$R_2 = 10K$

$$\text{base voltage} = \frac{12 \times 1.5}{11.5}$$
$$= 1.56 \text{ V}$$

If supply voltage = 12 V

$R_3 = 4K7$

$$I_c = \frac{6}{4.7} = 1.27 \text{ mA}$$

$V_e = 1.56 - 0.6$
$ = 1 \text{ V}$

$R_4 = 1.27K$
use 1K2

100 Hz as lowest frequency means that C_3 should be more than 2 µF. Would use 2µ2 or 4µ7.

Figure 12.8 *A better way of biasing a bipolar transistor for small-signal voltage gain. The method allows approximations to be used, but results in bias voltages that are very close to calculated values and which are very stable*

Configurations

The common-emitter or common-source type of circuit, or configuration, is by far the most useful of the ways in which a single transistor can be connected. It will achieve large figures of voltage gain *and* current gain (the ratio of (current signal out)/(current signal in)), with moderate values of input and output resistance. The output resistance of the stage is about equal to the load resistance R_3, and the input resistance is the resistance of the base-emitter diode, which is non-ohmic. This non-ohmic resistance is the main source of distortion when a bipolar transistor is used as an amplifier, because it ensures that the base signal current is never exactly proportional to base signal voltage. This **non-linear distortion** is reasonably low when the signal amplitudes are very small, and can be reduced further by using negative feedback.

The transistor can be used in two other configurations, common collector and common base – the parallels for the FET are common-drain and common gate. The **common-collector** configuration is also called the **emitter-follower**, and a practical example is shown in Figure 12.9. The collector is connected directly to the supply, so that *as far as AC signals are concerned* it is earthed and therefore common to input and output. The load resistor R_3 is now in the emitter circuit, and since the transistor maintains an almost constant voltage between base and emitter, the variation of voltage at the emitter will be almost identical to the variation at the base. The voltage gain of this circuit is therefore always less than unity, and a good working assumption is around 0.95. The output is in phase with the input, and there will be current gain. The importance of the circuit is that the output resistance is very low: it is as if the output signal were coming from a generator with only a few ohms of resistance. At the same time, the input resistance is high, higher than you would expect from the diode action of base and emitter. This is a circuit with 100 per cent negative feedback, since the signal between the base and the emitter is equal to input signal – output signal, and therefore there is low distortion. The emitter-follower type of configuration is used extensively when a resistor is needed to supply signal to a load with low resistance, high capacitance or both.

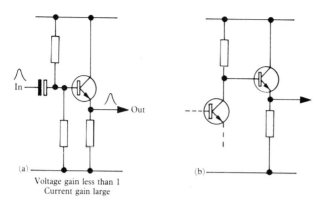

Voltage gain less than 1
Current gain large

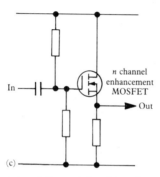

Figure 12.9 *The emitter-follower (a) and its advantages. A common method of ensuring correct biasing of the emitter follower is to connect the base directly to the collector of a previous stage (b). The MOSFET version is shown in (c)*

A development of the common-emitter circuit is the compound emitter-follower, or the **Darlington circuit** (Figure 12.10). A Darlington arrangement uses two transistors with the emitter of one connected directly to the base of the next, with no resistors connected to this point. This arrangement, which can be obtained in a single transistor casing, allows very high current gain, very high input resistance and fairly high voltage gain. A FET equivalent can also be obtained in a single casing.

That leaves the **common-base** configuration, illustrated in Figure 12.11. The common-base circuit has a very low input resistance, a high output resistance, and a fairly small voltage gain. Its main advantage is that it can cope with signals at high

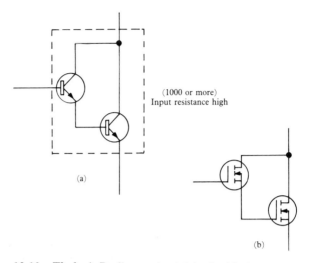

Figure 12.10 *The basic Darlington circuit (a) using bipolar transistors. The arrangement can be obtained in one package which can be used as an emitter-follower or as a common-emitter amplifier. The FET equivalent (b) is used only for a few special purposes*

frequencies better than the other configurations, and so its main use is in amplification of very high frequency signals. It is used in the early stages of FM radio or TV receivers, or in transmitter circuits, where the frequency is particularly high. In most examples, only one common-base transistor will appear in such circuits, because the superhet system changes the frequency of the incoming signal to a lower (IF) value that can be handled by transistors connected in the common-emitter configuration. The output of the common-base amplifier is in phase with the input, so that some care has to be taken to avoid feedback from output to input since this would cause oscillation. For some purposes when very high frequency oscillations are needed, a capacitor can be connected between the collector and the emitter of a common-base tuned amplifier circuit.

Circuit types

The types of circuits that can be constructed using transistors are classed as untuned amplifiers, tuned amplifiers, oscillators and switching (digital) circuits. Each type of circuit can be constructed

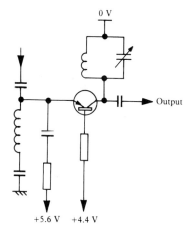

Figure 12.11 *The common-base circuit, used here as an RF amplifier for a high-frequency carrier. The FET equivalent, the common-gate amplifier, is seldom used*

using one single transistor stage, but it is much more common to find several stages used. In most modern equipment, separate (discrete) transistors are now used only for specialised purposes, and the bulk of a circuit will be constructed from integrated circuits. ICs are dealt with in detail in the two following chapters, so that this chapter will continue to deal mainly with the basic types of circuits as applied to single transistor stages.

Untuned amplifiers

The basic circuit for an **untuned** (or **aperiodic**) single transistor amplifier is that of the common-emitter stage illustrated in Figure 12.8. This is a typical small-signal Class A stage. The **Class A** term indicates that the biasing is arranged so that the transistor conducts at all times over the whole of the signal cycle. **Classes B** and **C** refer to bias conditions in which the transistor may be cut off for part of the cycle, and these arrangements are useable only when the rest of the signal can supplied in some other way, such as by using another transistor for part of the signal or by using a tuned circuit to supply the missing part of the cycle. Class A stages like this can be connected together (**cascaded**) so as to obtain greatly increased gain, and the overall gain of such stages is equal to the product of the gains of each stage. The gain can be so large that it is normal to include some negative feedback over all the stages of the amplifier. This not only reduces the gain, but will control and stabilise the amount of gain, allowing amplifiers of predictable and reproducible performance to be constructed.

For most purposes, the gain of such amplifiers will be quoted in **decibels** (dB). This is a logarithmic scale which is better suited for expressing gain for several reasons that are mostly tied up with the way that gain affects our senses. Suppose, for example, that we have an amplifier used in a radio receiver and that we are listening to a sound of constant loudness. Increasing the gain of the amplifier will make the sound appear to be louder, but not in the ratio of the gain. Doubling the gain of the amplifier, for example, makes the sound seem louder, but certainly not twice as loud. This was recognised by the telephone pioneer Alexander Graham Bell (whose invention was intended to be used as a deaf-aid), and so the unit of loudness was named the 'bel' in recognition of this. The bel unit is too large, however, and one tenth of a bel, a decibel, is more useful. The definition of a decibel is in terms of power ratio, equal to the logarithm of the ratio of two powers, usually output power and input power, Figure 12.12(a). This is the only correct definition of the decibel, but for comparing voltage gains, the formula in Figure 12.12(b) is used. The idea behind this is that power is proportional to voltage squared, and in logarithmic terms, squaring is carried out by multiplying a logarithm by 2, making the factor 20 in place of the 10 that is used for the power decibel.

(a) Power ratio in dB is $10 \times \log\left(\dfrac{P_2}{P_1}\right)$

where P_2, P_1 are power levels of signal
For example, a power gain of 100 gives 20 dB

(b) Voltage ratio in dB is $20 \times \log\left(\dfrac{V_2}{V_1}\right)$

where V_2, V_1 are voltage levels of signal
For example, a voltage gain of 100 gives 40 dB.

Figure 12.12 *Decibels: the correct definition (a) is as a ratio of two power levels. Note that figures such as 50 dB imply ratios only and are meaningless unless you know what is being compared. The 'voltage decibel' is often used (b) to compare voltage gains, but this is valid only if the impedance levels are identical*

Power output

A power output stage must control the amount of power in a load, so that the transistor needs to be able to provide signals that consist both of large voltage amplitude and large current amplitude. The power dissipated in the load will then be equal to $(V_L \times I_L)/8$, where V_L is the peak-to-peak voltage amplitude, and I_L is the peak-to-peak current amplitude of the signal across the load, assuming that this is a sine-wave. Most of the loads that are encountered consist of low values of resistance, so that transistor power amplifiers are required to pass currents of several amps AC through the load, using voltage amplitudes of typically 20–70 V.

Most modern power output stages use the transistors to supply the current directly (as opposed to the use of a transformer), and a typical circuit is shown in outline in Figure 12.13. This uses two bipolar transistors as emitter-followers in series, one *p-n-p* and the other *n-p-n*. If we ignore bias for the moment, we can see that the signal that is applied to the two bases will affect the transistors in opposite senses. When the drive signal is on its positive half-cycle, the upper transistor *Tr*1 will be switched on, because it is an *n-p-n* type. Since the circuit is wired like an emitter-follower, we can expect the waveform at the emitter of this transistor to follow the waveform at the base. On the negative half-cycle of the input, the *n-p-n* transistor *Tr*1 is cut-off and this time the *p-n-p* transistor *Tr*2 behaves like an emitter-follower, producing an accurate replica of the negative half of the input waveform at its

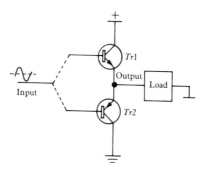

Figure 12.13 *The principle of the complementary output stage. This uses a p-n-p/n-p-n pair wired as a double emitter-follower. No bias arrangements are shown*

emitter. The circuit therefore splits up the signal handling between the two transistors so that each handles half a cycle.

In practice, it is not quite so simple. Unless the transistors are suitably biased, there will be a region in which neither conducts, and in order to avoid very large signal distortion, both transistors should be passing current at small signal amplitudes. Also, some method has to be found to allow the amplitude of the input signal to rise to a level higher than the voltage that supplies the output stage, because the base of $Tr1$ must be at least 0.6 V higher than its emitter voltage if $Tr1$ is to remain conducting. One common solution is shown in Figure 12.14, in which the input signal is **bootstrapped**. The output signal is fed back, in phase, to the point marked Y between the bias resistors for the base. This feedback will ensure that as the output voltage rises, the voltage available at the base will also rise, preventing cut-off. There should be no risk of oscillation, because the voltage gain of an emitter follower is always less than unity.

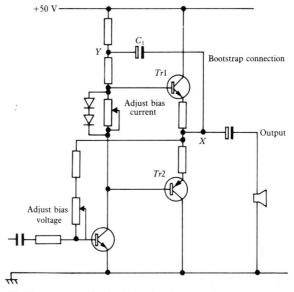

Figure 12.14 *A more detailed circuit of a complementary output stage, showing feedback, driving and bias arrangements. The signal at X is fed back through C_1 to point Y, so allowing Tr1 to conduct when the voltage at X is high*

The advantages of this type of stage, called the **single-ended push-pull** or **totem-pole** circuit, are high efficiency, ability to operate at high frequencies, and absence of inductors. The transistors need not be operated with either cut-off at any time, but this is the most common application of the circuit. If the transistors are biased so that both conduct at all times, in Class A, then each transistor will dissipate power throughout a cycle. For example, suppose the supply voltage is 50 V, and the steady bias current is 1 A, then the power dissipated is 50 W, shared between the two transistors.

This will not change when signal is applied, but the maximum possible signal is one that will swing the voltage at the output between 0 and +50 V, and the current between 0 and 2 A. This leads to an output to the load of $50 \times 2/8 = 12.5$ W for a sine wave. The maximum possible output power, in other words, is one quarter of the power dissipated as heat in the transistors, so that the efficiency is 25 per cent. When the stage is operated in Class B, with each transistor cut off for half of each cycle, the efficiency is much higher so that for a given output power, the dissipation in the transistors is much lower. The efficiency can be as high as 75 per cent, with the added bonus that the transistors dissipate power only when the load is being supplied with power, whereas in Class A biasing the dissipation is constant whether power is being delivered or not. The distortion of the Class B type of bias is higher, but this can be overcome by using negative feedback. Because of the limited safe operating area of bipolar transistors, the Class B biasing is preferred for bipolars, though Class A can be used safely with power FETs.

Tuned amplifiers

A **tuned amplifier** is an amplifier whose gain is a maximum for some particular frequency and much lower for other frequencies. All tuned amplifiers will have a measurable bandwidth, meaning the range of frequencies around the tuned frequency for which the gain is at least 70 per cent of the maximum gain (see Figure 9.16). The frequency to which an amplifier is tuned is usually referred to as the **centre frequency**, so that the bandwidth is the range of frequency from the centre frequency to the frequency at which the gain is 70 per cent of the gain at the centre frequency.

The graph of gain plotted against frequency is usually symmetrical, so that the bandwidth is the same on either side of the centre frequency. If this is not so, then the bandwidth has to be quoted as two separate figures.

A tuned amplifier must make use of a tuned circuit, which is usually the load of the amplifier. In some cases, the input circuit may be tuned as well, but the load is the important tuned circuit to consider for most purposes. A typical tuned load would be a parallel tuned circuit whose resistance is a maximum at the frequency of resonance. Since the voltage gain of a transistor stage depends on the resistance of the load, this makes the gain a maximum at the frequency of resonance as is required.

Figure 12.15 shows a typical single transistor circuit, using a tuned load. The output signal from the load will usually be taken from a tapping on the coil, or from another coil wound on the same former, so as to provide a signal with comparatively low source resistance for the next stage. The biasing of this type of stage would depend on how it was to be used. For a receiver circuit, Class A biasing would be used, but for a transmitter in which this stage was being used as a power amplifier, the bias would be Class C. This implies that the transistor would conduct only on the peaks of the signal, and the rest of the waveform would be provided by the oscillation of the tuned circuit.

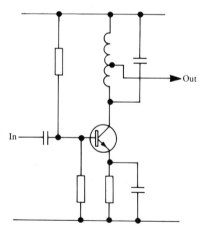

Figure 12.15 *A typical single tuned amplifier stage for carrier or IF frequency. It is now uncommon to find transistors used in this application because of the availability of ICs*

In tuned amplifiers for receiver circuits, as in so many other applications, it is normal to turn to ICs rather than to work with separate transistors. Op-amps can be used in tuned amplifiers, subject to the usual bandwidth and slew-rate restrictions, but it is more usual to employ special-purpose ICs for the particular range of frequencies that is being used. ICs are available for the usual range of radio and TV IF signals, and also for incoming radio frequencies in various ranges. These ICs do not include tuned circuits, and suitable tuning stages have to be connected to terminals on the IC. These tuning stages may be the parallel or series resonant type using inductors and capacitors, or they may consist of more modern devices such as surface-wave acoustic (SAW) filters. The **SAW filter** is formed on a quartz crystal, and consists of sets of metal electrodes. An input signal is applied between one pair of electrodes, and this causes the crystal to vibrate mechanically at the frequency of the signal. By grinding the crystal to a suitable thickness, and using a specified shape and pattern of metal electrodes, the vibration can have a resonant peak at the signal centre frequency. Like many other types of mechanical vibrations, this vibration will have a very small bandwidth, much lower than can be achieved with circuits using inductors and capacitors. Another pair of electrodes on the crystal converts these mechanical vibrations back into electrical signals, providing an output. In most cases, a single SAW filter may be all that is needed for a complete tuned amplifier of many stages. For TV and radar purposes, it is important to have a wide bandwidth, so that inductor-capacitor tuned circuits with damping resistors are more useful.

Oscillators

An **oscillator** is a circuit that provides an AC output for a DC (power-supply) input. We can think of oscillators either as amplifiers with positive feedback, or as devices which simulate negative resistance, meaning that the voltage across the device drops when the current increases. In either case, some circuit components must be used to determine the frequency at which oscillation will take place. If the frequency-determining components consist of a tuned circuit (including quartz crystal, SAW and other circuits), then the oscillator will deliver sine waves at the frequency of resonance of the tuned circuit. If the frequency-determining circuit consists of capacitors charged and discharged through resistors, the waveform is much more likely to be square in shape, and this type of oscillator is sometimes called **aperiodic** or a **relaxation oscillator**.

The simple tuned-circuit oscillator is typified by two ancient designs, the **Hartley** and the **Colpitts** oscillators. One version of each of these types is illustrated in Figure 12.16, and the feature that they have in common is that part of the output signal is tapped to be used to provide positive feedback. In the Hartley oscillator, the tapping is of an inductor; in the Colpitts oscillator the tapping is provided by using a potential divider made from capacitors, but the principle is the same. The voltage tapped from the capacitors in the Colpitts circuit is in phase with the output signal, so that for positive feedback it has to be connected to the emitter of the single transistor. The Hartley oscillator coil can be designed so that the tapping produces a feedback voltage that is inverted, allowing this to be connected to the base of the transistor. The amount of feedback is fairly critical. If too little feedback is used, the circuit will not oscillate, and if too much feedback is used the waveform will be poor, not a perfect sine wave by any stretch of the imagination. By varying the bias current through the transistor, it is usually possible to obtain a setting that will just allow reliable oscillation with a good waveshape, and this type of control is easier than trying to control the amount of feedback.

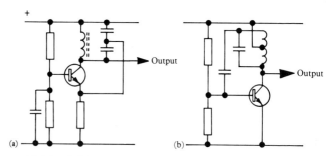

Figure 12.16 *Colpitts (a) and Hartley (b) oscillator circuits. The outputs would normally be taken from a tapping on the coil, from a small additional winding, or from a buffer stage like an emitter-follower so as to reduce the load on the oscillator*

The Colpitts and Hartley oscillators, though widely used, are not the only types of transistor oscillator. Other types include the **double-tuned oscillator** and the **common-base oscillator**, as illustrated in Figure 12.17. The drawback of all oscillators that make use of inductor-capacitor circuits, however, is that the waveform is never a perfect sine wave. This is because no circuit of this type can ever have a narrow enough bandwidth to

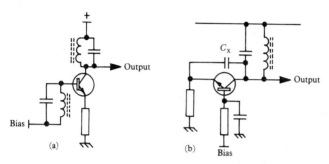

Figure 12.17 *Two other oscillator circuits, the tuned-base tuned-collector (a) and the common-base type (b)*

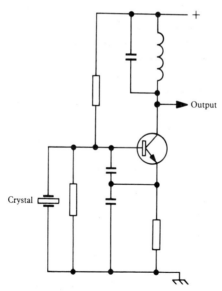

Figure 12.18 *A typical crystal-controlled oscillator stage – this is of the Colpitts family*

exclude completely harmonics of the oscillating frequency. Much better quality sine waves are obtained if a quartz crystal is used as the resonant circuit, as illustrated in Figure 12.18. The improvement is measured by a quantity called the **Q-factor**, which can be around 100 for capacitor-inductor circuits, but 30 000 or more for a quartz crystal. The other advantage of using a quartz crystal is frequency stability. An oscillator that is crystal controlled can be relied on to keep its frequency constant to within a few Hertz, particularly if the crystal is maintained at a constant temperature (in a crystal oven). Such stability is essential for all types of radio frequency transmitters and for applications in receivers that call for a precise frequency, such as colour TV reception and narrow-band radio communications. Crystals are also used to determine the frequency of oscillators that do not deliver sine waves, and are used in such applications (mainly in computers) because a precisely controlled frequency is needed for timing purposes.

Switching and relaxation oscillators

So far, we have looked on the transistor as a linear device, i.e. as used in applications in which the graph of output plotted against input is a reasonably straight line. An alternative use for the transistor, for which it is in many ways better suited, is **switching**. Switching means that the collector-emitter circuit of the transistor will either be off, passing no current, or on, passing as much current as the resistance of the rest of the circuit will permit. The very large current gain of the bipolar transistor makes this action fairly easy, because at a base voltage of just below 0.6 V (for a silicon transistor) the current will be zero, but for a base voltage of very little more, typically 0.65 V, the collector-emitter current will be large. Switching can therefore be easily accomplished by altering the base voltage, though in order to avoid excessive current through the base it can be necessary to connect a resistor in series. The FET is even easier to arrange as a switch, because it is necessary only to provide for a voltage signal, with no current required at the gate other than the charge and discharge capacitances.

Figure 12.19 shows a typical switching stage for a bipolar transistor. A square pulse is applied to the base terminal through a resistor which will limit the base current to a safe value. The collector circuit uses a small load resistor, so that the waveform at the collector is an inverse copy of the input wave. In many switching circuits, particularly digital circuits, there is no amplification as such, because all waveforms have about the same amplitude. The emphasis of switching circuits is on speed of response, so that there is only a negligible time from the start of the voltage change at the base to the start of the voltage change at the collector, for example. This means that any stray capacitances in the circuit must be charged or discharged as rapidly as possible, and this is accomplished by using large current values in the collector circuit. In the simple switching circuit of Figure 12.19, the drop in voltage at the collector can be very rapid when the transistor is switched on, but the rise in voltage when the transistor is switched off is not so fast, because the stray capacitances have to discharge through the load resistor.

Faster charging and discharging can be achieved by using circuits that make use of more than one transistor, and a favourite method is illustrated in Figure 12.20. The *n-p-n* transistor will switch on when the input voltage is high, and the *p-n-p* transistor will switch off at the same time. This allows the current flowing through the *n-p-n* transistor to discharge any stray capacitances in the output circuit. When the input voltage drops again, the *p-n-p*

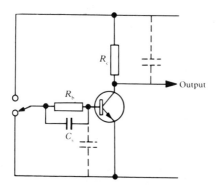

Figure 12.19 *An elementary switching circuit using a bipolar transistor. The resistor R_b is needed to limit the size of the current that will flow when the transistor is switched on, and capacitor C_s compensates for the stray capacitance at the base, shown dotted. The rate of voltage rise when the transistor is switched off is determined by how fast the stray capacitance at the collector can be charged through R_c*

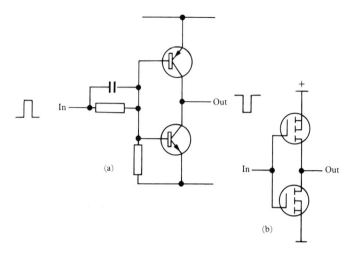

Figure 12.20 *Using a complementary* p-n-p n-p-n *pair (a) for switching. This results in faster switching in both directions because more current is available to charge and discharge stray capacitance in each direction. The version shown is an inverter, but the positions of the transistors can be reversed to achieve a non-inverting switch. The FET version (b) is also shown*

transistor switches on and the *n-p-n* transistor switches off, with the current through the *p-n-p* transistor charging the stray capacitances rapidly. By making every switch stage of this double-acting type, the switching speed for both on and off directions can be greatly improved. The FET version of the circuit is also shown – its action is virtually identical. From now on, the diagrams will show both the bipolar and the FET versions if there are significant differences, but the description will refer to the bipolar version, since the bipolar version is generally the more difficult to work with.

The astable oscillator

The **astable oscillator,** or **multivibrator,** is an oscillator that makes use of switching, and uses capacitance charging and discharging to fix the times of each part of the cycle. A typical circuit is illustrated in Figure 12.21, along with the waveforms for one transistor. The times for the main portions of the cycle are determined by the time constants of $C_1 R_2$ and $C_2 R_3$, and the

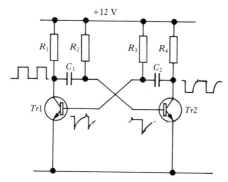

Figure 12.21 *An astable multivibrator and its waveforms. The negative-going edge of each pulse at the collector is fast, but the positive going edge can be slow because stray capacitance has to be charged through the resistor R_1 or R_4. The FET version is almost identical*

times of switching are determined by the size of stray capacitance and the current that each transistor can pass.

Imagine that $Tr1$ has just been switched on because the voltage at its base has risen to the level that just allows current to pass. Any drop in voltage at the collector of $Tr1$ will make the voltage at the base of $Tr2$ drop, switching off $Tr2$ and so causing the voltage at the collector of $Tr2$ to rise. This in turn will raise the base voltage of $Tr1$, completing a positive feedback loop in which there is a very large figure of gain. Because of this positive feedback loop, the changeover from conducting to non-conducting, or vice versa, is always very fast, and we can concentrate on what happens between these times, the relaxation time.

When $Tr1$ is conducting fully and $Tr2$ is off, the effect of the negative pulse at the collector of $Tr1$ will have been to make the base voltage of $Tr2$ go negative, roughly to -12 V if the supply voltage is $+12$ V. One plate of C_1 is held by the collector of $Tr1$ at about earth voltage, and the other plate is at about -12 V, but connected by resistor R_2 to $+12$ V. This capacitor will therefore discharge, and will take some time T_1 for the voltage at the base of $Tr2$ to reach about 0.6 V. At this voltage level, $Tr2$ will switch on and $Tr1$ will switch off. Capacitor C_2 is now charged, and will start to discharge through R_3, a processs which ends after some time T_2 when the base of $Tr1$ reaches the switch-on voltage, when the whole cycle repeats. The times T_1 and T_2 are approximately

70 per cent of the time constant values C_1R_2 and C_2R_3 respectively, and the time for one complete cycle is $T_1 + T_2$.

The waveform at the collector of each transistor is approximately square, and the two collectors provide waveforms that are inverted with respect to each other. The waveform at each base is part of a charging curve which for some purposes can be taken as approximately a sawtooth shape. In practice, the waveform at the collector is more useful, and a variety of circuits can be used to make the rising side of the wave of better shape so as to provide a good square wave.

If the circuit is constructed with one capacitor coupling and one resistive coupling, as in Figure 12.22, it becomes a **monostable** or **one-shot**. In this particular variety of the circuit, transistor $Tr2$ is normally conducting because its base is connected to the supply voltage through resistor R_2. The collector voltage of $Tr2$ will be very low, preventing $Tr1$ from conducting because of the connection through R_4 to the base of $Tr1$. A brief positive pulse (or **trigger** pulse) can be injected at the input, passing through the diode D_1 and causing $Tr1$ to conduct. When this happens, the positive feedback loop through C_1, $Tr2$ and R_4 will make the transistors switch over very rapidly, and $Tr2$ will switch off with its base voltage well below earth voltage. The capacitor C_1 will then discharge as before until $Tr2$ can conduct again, at which time $Tr1$ is rapidly switched off again. The output is therefore a

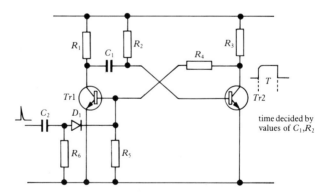

Figure 12.22 *The monostable. This is used to produce a square pulse of specified duration (pulse-length) from a brief trigger input. The time constant C_1R_2 determines the duration of the output pulse*

square pulse, whose duration is set by the time constant of C_1R_2, and whose leading edge coincides with the trigger pulse.

Finally, the circuit can be constructed with no time constants at all, as a **bistable** or **flip-flop** (Figure 12.23). This circuit, as the name suggests, will remain indefinitely with one transistor conducting and the other non-conducting, until an input is used to switch over to the alternative state. In the simple version of Figure 12.23, the switch over will be accomplished by a pulse on one or other inputs, but another version of the circuit can be made to **toggle**. Toggling means that a single pulse at one input can be used to make the circuit change over, and this is accomplished by the use of diodes called **catching diodes**.

The circuit is shown in Figure 12.24, and each change of output requires a negative pulse at the input. Imagine that $Tr1$ is conducting, so that its collector voltage is low and the base voltage is being held at about 0.6 V. At the same time, $Tr2$ will be off, with its base voltage held at nearly zero volts, and its collector voltage high, at about supply voltage. All of this means that diode D_1 is almost conducting, but diode D_2 is heavily reverse-biased. Now when a negative pulse of a volt or so is applied at the input, it will make diode D_1 conduct rather than diode D_2, because D_2 is so thoroughly biased off. The pulse will therefore switch off $Tr1$, causing the switchover. After the switchover, D_1 is reverse biased

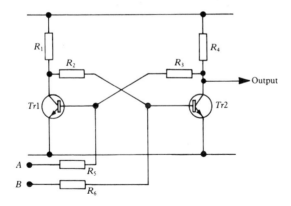

Figure 12.23 *The simple bistable or flip-flop circuit. At the moment of switching on, one or other transistor will conduct, keeping the other switched off until a pulse at input A or B reverses the states. Circuits like this are the basis of memory and registers for computers*

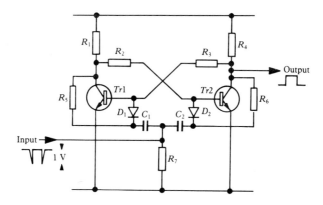

Figure 12.24 *A toggling bistable which will reverse state at each trigger pulse input. As with all circuits of this type, the output can be taken from either collector depending on whether a positive or a negative pulse is required*

and D_2 is almost conducting, so that the next negative pulse at the input will switch back to the first state. In such a circuit, there will be one complete pulse at a collector for two pulses at the input, so that the circuit is sometimes described as a **scale-of-two**.

13
Linear ICs

A transistor is manufactured by a set of operations on a thin slice, or **chip**, of silicon crystal. These operations include selective doping and insulation, and the areas that are affected can be controlled by the use of metal masks placed over the semiconductor. Also, by using various amounts of doping of a strip of semiconductor material, the resistance of the strip can be controlled, so that it is possible to fabricate a resistor on a semiconductor chip. In addition, because silicon oxide is an excellent insulator, it is possible to make capacitors by doping the semiconductor to make a connection, oxidising to create an insulating layer, and depositing metal or semiconductor over the insulation to form another connection. Since transistors, resistors and capacitors can all be formed on a silicon crystal chip, then, and connections between these components can also be formed, it is posssible to manufacture complete circuits on the surface of a chip. The transistor types can be bipolar or FETs, or a mixture of both, but FETs are preferred because they are easier to fabricate in very small sizes.

Integration

The advantages of forming a complete circuit, as compared to the discrete circuit in which separate components are connected to a printed circuit board (PCB), are very great. There is the convenience of having very small complete circuits. It is possible to pack huge numbers of components onto one chip simply by making the components very small, and the technology of creating very small components has advanced spectacularly. The idea of packing a million FET transistors onto one chip might have seemed unbelievable only a few years ago (and it is still hard to imagine even now), but this is the extent to which we have come. Thus we can manufacture very complicated circuits that simply would not be economic to make in discrete form.

Another advantage of integrated construction is its cost. The cost of making a circuit that contains 50 000 transistors is, after the tooling has been paid for, much the same as the cost of making a single transistor. This is the basis of the £2.50 calculator, the TV remote control, and the small computer. The cost of integrated circuits has been steadily reducing with each improvement in the technology, so that prices of computers go down even faster than the prices of cars go up. The greatest advantage from the use of integration, however, has been reliability, and it was the demand for 100 per cent reliability in the electronics for the space missions that fired the demand for ICs.

Consider a conventional circuit that uses 20 transistors, some 50 resistors and a few other components. Each transistor has three terminals and each resistor has two terminals, so that the circuit contains at least 160 soldered connections. One faulty connection or one faulty component will create a fault condition, and the greater the number of components and connections there are the more likely it is that a fault will develop. By contrast, any IC which will do the same work might have only four terminals, so that it is connected to a PCB at four points only.

The IC is a single component which can be tested – if it works satisfactorily it will be used. Its reliability should be at least as good as that of a single transistor, with the bonus that each internal part will be working under ideal conditions if the IC has been correctly designed. The reliability of a single component is 20 times better than the reliability of 20 components that depend on each other. If the IC does the work of a thousand components, its reliability will be about a thousand times better than that of a single component and so on. This enormous improvement in reliability, many times greater than any other advance in reliability ever made, is the main reason for the overwhelming use of ICs. When electronic equipment fails, the first thing to check is the connection to the mains plug, not the state of the ICs.

For these reasons, circuits that use discrete transistors are becoming a rarity in modern electronics, which is why comparatively few circuit examples have been shown using transistors. When we work with ICs, circuits become very different. We may still be concerned with DC bias voltages, but usually at just one terminal. We are still concerned with signal amplitudes and waveshapes, but only at the input and the output of the IC. What goes on inside the IC is a matter for the manufacturer, and we simply use it as advised. It is a complete circuit, and servicing amounts to deciding whether the IC is correctly dealing with the signal that is applied to its input. If it is not, then it has to be replaced. If the input and output signals are correct, the fault lies elsewhere. The use of ICs has not complicated servicing, nor has it made designers redundant. On the contrary, it has resulted in a huge increase in the amount of items to service and the number of goods that can be designed.

In this section, we shall be looking at linear ICs. These are the types of ICs that deal with waveforms, in which amplitude and waveshape of the signals at the input and at the output are important. The other main class of ICs are the digital ICs that deal with on/off signals, and we shall encounter them in the next chapter. For a number of reasons, working with linear ICs is considerably more difficult, and the range of circuit actions is very much wider.

Op-amps

For most purposes an IC of the type called the **operational amplifier** (usually shortened to **op-amp**) will be used in place of a multi-stage amplifier constructed from individual transistors. The internal circuit of such an IC is not necessarily known to the user, and the important point to recognise is that the gain and other features of the op-amp will be set by external components such as resistors. This is true also of most amplifiers that have been constructed by using individual transistors, but since the op-amp consists of one single component, it is very much easier to see which components control its action, since only these components are visible on the circuit board.

The 'typical' op-amp is the type 741, which is a very old design of bipolar op-amp but one whose principles are followed by most of the modern types also. Figure 13.1 shows the connections to, and the symbol for, an op-amp of this type – the triangular symbol is used for any op-amp, but some of the connections are not used for modern op-amp types. It is important to note that the op-amp has two signal inputs, marked with + and – signs. A signal at the + input will appear amplified at the output and in

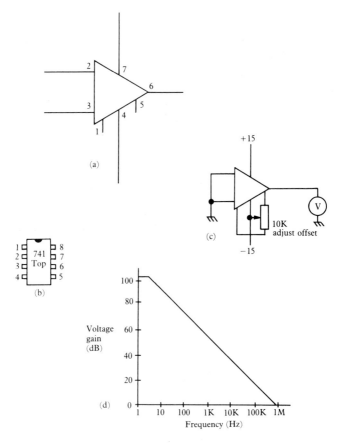

Figure 13.1 *The 741 op-amp, a most enduring design. (a) The symbol and pin diagram. (b) The chip layout – the 8-pin DIL package is the most common type. (c) Using the offset adjustment. The op-amp is used with equal voltage positive and negative supplies (balanced supplies), and with the inputs earthed the output voltage is set to zero by adjusting the offset potentiometer. This is needed only when the chip is used for specialised DC amplifier purposes. The gain-frequency graph is illustrated in (d)*

phase. A signal at the −input will also appear amplified at the output, with the same figure of gain, but inverted. If the same signal is applied to each input, then the output should be zero, because the action of the op-amp is to amplify the *difference* between the signals at its two inputs. If both inputs are used in this way, the amplifier is acting as a **differential amplifier**.

For many purposes, however, the op-amp is used single-ended, meaning that only one input signal is supplied consisting of a voltage whose amplitude varies with respect to earth. For such a signal, the op-amp can be used in two ways, as an **inverting amplifier** or as a **non-inverting amplifier**. The gain of the op-amp in its natural state is very high, of the order of 100 dB, equivalent to a voltage gain of 100 000 times, so that for almost all practical uses this gain has to be reduced by using negative feedback. The negative feedback connection is from the output terminal to the −input. If we simply connect a resistor from the output to the −input, then the negative feedback is 100 per cent: all of the output voltage signal is connected back to the input, and the gain will be unity. We need to be able to take a signal to the input, however, and if we want the output signal to be inverted relative to the input, then the signal input has to be to the − terminal. Connecting the input signal by way of a resistor (Figure 13.2) will

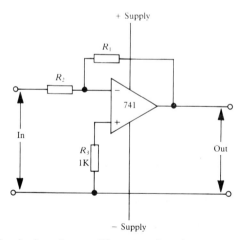

Figure 13.2 *An inverting amplifier connection of an op-amp. The gain is given by the ratio R_2R_1, and both input and output are at about DC earth levels because of the connection of the + input to earth through R_3*

reduce the amount of negative feedback, and so permit the gain to be greater than unity if required. In the circuit of Figure 13.2, the voltage gain is given by R_1/R_2 (in decibels, $20\log(R_1/R_2)$). To achieve unity gain, we can make these resistor values equal, and by using different values we can achieve various figures of voltage gain. The value of R_2 includes any circuit resistance of the source of the signal, and more easily predictable results are obtained if this resistor is of several K in value.

The DC bias of the op-amp is achieved most simply by using two supplies of equal positive and negative voltages, typically $+12$ V and -12 V. When this is done, then provided that there is a resistor of not too high a value connecting the output with the $-$input, the op-amp will be correctly biased. If it is inconvenient to have to supply voltages like this, the circuit can be rearranged to the type shown in Figure 13.3, using a single voltage supply. In this case, however, capacitors are needed to isolate the DC voltages. The DC voltage on the $+$ input is set to half of the supply voltage by using resistors R_3, R_4 of equal value, but the remainder of the circuit is constructed in the same way as before,

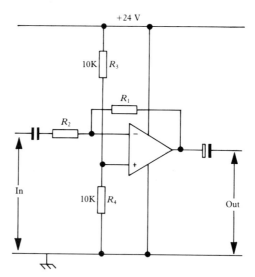

Figure 13.3 *Using an op-amp inverting amplifier with an unbalanced (single-ended) supply. Resistors R_3 and R_4 set the voltage at the + input at half of supply voltage, and the negative feedback through R_1 ensures that the output voltage will be almost exactly the same*

with capacitors now used to feed signal in and out. The DC level of both input and output will be approximately half of the supply voltage.

The inverting amplifier connection of the op-amp is used to a large extent, but for some purposes a non-inverting connection can be useful. The amplifier is non-inverting if the signal is taken to the + input, but the feedback must still be connected to the −input, and the amount of feedback will decide how much gain is obtained. Figure 13.4 illustrates a typical non-inverting circuit, using balanced + and − power supplies. The voltage gain is now equal to $(R_1+R_2)/R_2$, so that if unity gain is needed, the resistor R_2 should be omitted. The input of signal is taken to the + input terminal, with a resistor connected to earth to ensure correct DC conditions as usual. The input resistance of the op-amp is very high, but the use of R_3 makes the value equal approximately to the size of this resistor. The output resistance, as always, is low, a few ohms. The version of this circuit for a single-ended power supply is illustrated in Figure 13.5.

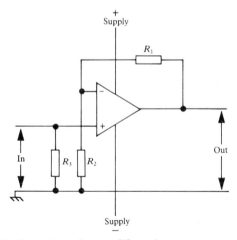

Figure 13.4 *A non-inverting amplifier using an op-amp with balanced power supplies. The gain is set by* R_1 *and* R_2, *and the value of* R_3 *can be large, about 1M*

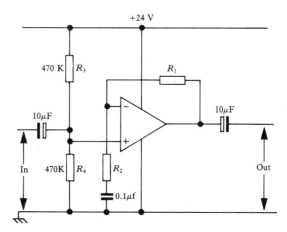

Figure 13.5 *The single-ended supply version of the non-inverting amplifier*

Slew and Bandwidth

The use of op-amps does not solve all the problems of providing gain, though for low-frequency signals there are very few problems to overcome. Because the op-amp is a very complicated circuit that makes use of many transistors in common-emitter circuits, its gain for high-frequency signals will be lower than its gain for low-frequency signals. More seriously, its gain for large amplitude fast-changing signals will be lower than for low-amplitude signals of the same frequency. This latter problem is one of **slew rate**.

The slew rate of a signal is the rate of change of voltage in the signal, in terms of volts per second. In practice, this gives figures that are much too large, and the more practical unit of volts per microsecond is used. Two waves of identical frequency and waveshape can have very different slew rates, depending on their peak amplitudes. Suppose, for example that we have a wave which is of sawtooth shape and which repeats at 100 kHz. It is easy to design an op-amp circuit that will provide a large voltage gain for a wave with this frequency, but the slew rate also has to be considered as Figure 13.6 shows. If the amplitude of the wave is 1 V, then for a time of 10 μs (corresponding to a frequency of 100 kHz), the slew rate is 1 V in 10 μs, equal to 0.1 V/μs. For a 10 V amplitude, however, the slew rate is 10 V/10 μs, equal to

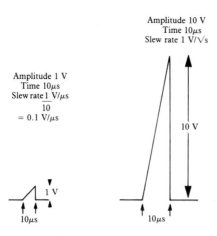

Figure 13.6 *How the amplitude of a wave affects slew rate. A sawtooth wave has been chosen because it is the easiest to use for calculations*

1 V/μs. The amplifier must cope with the slew rate of the highest amplitude of signal at the output. As it happens, many op-amps will cope with a slew rate of 1 V/μs, but the 741 type is limited to a slew rate of 0.5 V/μs.

For higher slew rates, however, it may be necessary to specify different op-amps. The maximum slew rate cannot be altered by circuit tricks; it is fixed by the internal design of the op-amp itself. Unlike the bandwidth for small sine wave signals, which can be increased by using more negative feedback and sacrificing gain, slew rate is a fundamental figure for the op-amp. High slew rate op-amps can be obtained for applications in which this figure must be large. To take an example, the RS 5539 op-amp has a quoted slew rate of 600 V/μs, making this op-amp suitable for very fast-changing signals, including sine waves approaching the 1 GHz range. This, however, is exceptional, and slew rate figures in the range 0.5–10 are more common.

The slew rate is a very useful figure to use when an op-amp is employed in circuits that use sawtooth, pulse and other non-sine waveshapes. For sine wave use, an alternative figure, the **power bandwidth** is more useful. The power bandwidth is the maximum frequency of signal that can be amplified at full power. If higher frequencies are used, then lower gains must be used so that the product of gain in decibels and frequency in (usually) MHz or kHz is the same. In other words, if the gain of an op-amp is quoted as 100 dB (gain of 100 000 times) and its power bandwidth is 10 kHz, then you can expect to be able to use the full 100 dB of gain on signals up to 10 kHz, and if you make the gain 10 000 (80 dB), you can use signal frequencies up to 100 kHz, and so on. To put it another way, each 20 dB reduction of gain (by using negative feedback) will give a tenfold rise in maximum operating frequency. This, however, is still subject to the slew rate limitation, so that the slew rate should always be calculated first. For a sine wave, the slew rate in V/μs is $6.3 \times V_0 \times f$, where V_0 is the peak amplitude, and f is the frequency in MHz. For example, a sine wave of 5 V peak and 0.5 MHz frequency would have a slew rate of $6.3 \times 5 \times 0.5 = 15.75$ V/μs. Only a limited number of op-amp types can cope with this slew rate, though many could give the required bandwidth for signals of lower amplitudes.

CDA and transconductance amplifiers

Another type of op-amp that is used for some purposes is the **current-difference amplifier** (CDA), otherwise known as a **Norton amplifier**. In an amplifier of this type, the output voltage is proportional to the difference between the *currents* at the two inputs. This leads to circuits which look very odd when compared to the usual type of op-amp, because the input currents are set with the aid of large-value resistors, as Figure 13.7 illustrates. The current to the + input is set with a 2M2 resistor, and since the voltage at the output is normally required to be about half of the supply voltage, a 1M resistor from this point to the − input will set the bias current correctly. The advantage of using a CDA is that the voltage swing at the output can be very close to the limits of the supply voltage, making this type of amplifier useful in interfaces to digital equipment. Another comparatively rare type of op-amp is the **transconductance amplifier**, in which the transconductance (milliamps of current output per unit voltage difference between the inputs) can be chosen by setting the bias current. Such amplifiers are used in gain-controlled circuits.

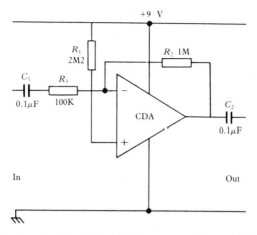

Figure 13.7 *A typical circuit for a CDA op-amp such as the LM3900. The output voltage is proportional to the difference between the input currents, and this requires large resistor values to be used to set the currents*

Specialised amplifiers

Operational amplifiers are the general-purpose amplifier types that are intended mainly for instrumentation use and in particular for use at DC and low frequencies. For audio and video signal amplification, more specialised ICs are available that are more suited to these particular purposes. Many of these ICs incorporate features which would be impossibly expensive to provide in discrete form unless the circuits were being provided on a build-to-order basis. For example, it is possible to buy microphone preamplifiers with VOGAD (**voice operated gain adjusting device**) incorporated. This IC, used as a microphone preamplifier, provides an almost constant 90 mV signal output while the microphone is being used. The IC cost is under £2, but it provides an action that was previously available only in broadcasting studios where speech quality was pursued regardless of price. This is an excellent example of an application for which a much greater demand exists when an IC can be provided at a reasonable price.

The less specialised audio ICs consist of **preamplifiers** and **power amplifiers**. The preamplifiers generally allow two stereo channels, with inputs for magnetic cartridge and tape at least, some feature very many other inputs. The specialised design ensures very low distortion and excellent channel separation, along with low noise. ICs of this type have almost completely replaced discrete transistor circuits even for equipment of the highest quality; the true hi-fi market was the last to accept ICs. The enormous advantage of ICs in this respect is consistency, because the efforts that were needed to ensure consistent high performance from discrete amplifiers were uneconomic.

Power amplifiers in IC form have made very considerable inroads on audio equipment, but have not yet captured the upper end of the hi-fi market. Nevertheless, the trend is to use the IC units, particularly in view of the kind of performance that can now be obtained. A power stage which can deliver 120 W into 8 ohms with a total harmonic distortion of 0.005 per cent at 1 kHz, signal-to-noise ratio of 100 dB, bandwidth of 15 Hz–100 kHz and input sensitivity of 500 mV for full rated output would not so long ago have been considered quite exceptional for a discrete-transistor stage. It is now available as a mass-produced IC that is built into a heat-sink ready for use, and it would not be easy to justify the construction of a discrete circuit when ICs of this specification are available. There are, of course, many audio power output ICs that provide more modest output powers (and at much lower prices) for a lot of the more mundane audio applications.

Some audio applications are more specialised, but are important and not easy to provide in discrete form. A typical example is companding. A **compander** (compressor-expander) is a device that is used for sound equipment in which signal-to-noise ratio is a problem, particularly in high-quality cassette recording. The principle is that of compression on recording and expansion on replay. During recording, the amplitude range of the signal is reduced, so that the volume difference between soft and loud is made smaller. This allows the signals to be recorded on cassette at a level that is well above the noise level, but without the risk of causing distortion because of excessive signal amplitude. On replay, the signal is expanded, making the soft passages softer, and the loud passages louder. This process is the exact reversal of compression, so that the music content of the signal should be exactly as it was before the compression process. The tape noise, however, which was at a lower level than the lowest level of the signal, will have been reduced to inaudibility. The provision of a compander in IC form solves the problems of matching the two actions, and is an excellent low-cost solution for problems of signal-to-noise in audio equipment.

Other audio ICs include linear attenuators, which allow full volume control of a signal to be achieved by a control voltage. Attenuators can also be digitally controlled, using a six-bit code (see Chapter 14) to control the attenuation level. Audio signal delay ICs can be used to provide delays for use in public address equipment, where the sound from a nearby loudspeaker has to be delayed so that it arrives at the ear at the same time as the sound from a distant loudspeaker. ICs are also available for filtering, amplitude modulation, and voltage-to-frequency conversion. The phase-locked loop is also available as an IC that is widely used in conversion from frequency to voltage, both in FM receivers and in computing circuits.

Video circuits

Another application for linear ICs is video amplification for TV, VCR and oscilloscope use. Video amplifiers, being pulse amplifiers, are also used in digital communications equipment and in controllers for magnetic memories. Video amplifiers demand very large bandwidths and much higher slew rates than audio amplifiers, so that the design and construction of good video amplifiers was always a more specialised business. ICs for video amplification offer a full range of facilities for low-level signals, though video output stages are still generally discrete. A typical video amplifier IC can offer gain figures up to 400 times, with rise times of a few nanoseconds and corresponding bandwidths of up to 90 MHz. Several buffer ICs can also be obtained for driving cables and other capacitive loads. Such ICs have no voltage gain, but can supply signal currents of up to 10 mA peak into a 1K load with very high slew rates (typically 1400 V/µs) and with bandwidths up to 125 MHz.

In addition, many of the other requirements of a TV receiver can be provided by the use of ICs. Most of these ICs are very specialised, often custom designed and produced for a specific receiver manufacturer. ICs exist for the IF stages, sound stages, video signal decoding, oscillators and synchronisation, so that very few functions of a colour TV receiver require the use of discrete transistors apart from the video outputs and the colour CRT itself. ICs are also available for other applications, such as modulation of video signals on to UHF carriers so that signals from video games and from computers can be connected safely into standard TV receivers. The use of these ICs is declining now that most computers provide for the use of a purpose-built monitor rather than the less satisfactory use of a domestic TV receiver.

Voltage regulation

Another class of circuits in which discrete transistors have now almost completely disappeared is voltage regulation. The purpose of a voltage regulator is to provide a DC output voltage which is fixed despite changes in output current and input voltage. To make this possible, the input voltage needs to be higher than the regulated output – at least 7 V for a 5 V regulated supply, for example. Because of the difference between input and output voltage levels, voltage regulators will dissipate power when current is drawn from them, and the amount of power dissipated is equal to the product of current and voltage difference. Voltage regulator ICs are therefore packaged like power transistors with a flat surface or tab which can be bolted to a heat sink (see Chapter 15). Fixed voltage regulator ICs are available for all of the commonly used voltage levels of 5 V, 12 V, 15 V and 24 V, and the ICs can be obtained in positive voltage or negative voltage types. If the regulated voltage supply is not one of these, or if the output must be varied, then ICs for variable voltage supplies are available. Greater care must be taken with heat dissipation for these variable supplies, because the voltage input has to be suitable for the highest output that will be used, so that there will be a comparatively large voltage across the regulator when a low output voltage is selected. The variation of voltage is achieved by using a third terminal connected to a potentiometer (Figure 13.8). Regulators can also be obtained for complementary supplies, meaning positive and negative supplies of equal voltage, as required for many op-amps.

The use of ICs in voltage regulation is not confined to DC voltages, and ICs for thyristor and triac control are also available.

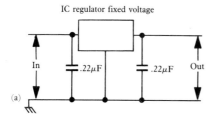

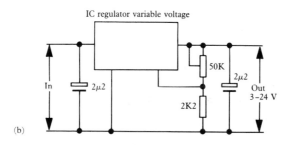

Figure 13.8 *Typical IC voltage regulator circuits for a fixed output voltage (a) and for variable output (b). The input voltage must be adequate to sustain whatever output is needed. The capacitors at the input and output terminals are used to suppress oscillations and should be wired as close to the pins of the IC as possible*

These controllers can be obtained in several forms, providing the two main methods of operation. The most straightforward method is **phase control** (see Figure 11.18), in which the switch-on of the thyristor or triac is varied by controlling the phase of a trigger signal with respect to the AC supply. An IC for this purpose, the TDA2085A, exists which can use the AC mains as its power supply and which provides a +5 V DC output for use by other circuits, as well as the trigger pulse for the thyristor or triac. When this method is used, the supply line must include series inductors to avoid the radiation of pulses which could result in false triggering of other thyristor equipment.

The alternative method (Figure 13.9), involves **zero-voltage switching**, and is suitable only for purposes in which the regulation can consist of supplying complete groups of cycles of AC. Equipment that makes use of heaters, motors with very large flywheels, pumps to supply reservoirs, etc., will have a large time

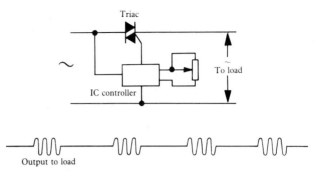

Figure 13.9 *Outline of a zero-voltage switch burst controller for AC. The AC is switched on or off at the zero-voltage point in a cycle for as many cycles on and off as are determined by the potentiometer setting*

constant. For such supplies, regulation can consist of supplying several complete cycles of AC followed by a gap, rather than by a continuous supply of part-cycles as is used in the phase-control method. This type of control system can use a small thyristor or triac to control a very large load by making use of zero-voltage switching, meaning that the device is switched on only while the AC voltage is zero. By using synchronised switching like this, there is none of the interference that would be caused by suddenly switching on a supply.

Regulation can be achieved in a third way, by **switch-mode** **power supplies**, (Figure 13.10). A switch-mode supply makes use

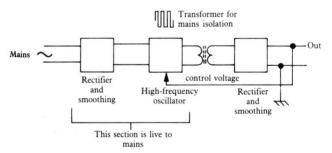

Figure 13.10 *Outline of a switch-mode power supply. The incoming AC is rectified and used to operate a high frequency square-wave oscillator that drives an output stage. The output is rectified and smoothed, requiring only small capacitances, and used to provide DC. This is stabilised by using the output DC level to control the oscillator. A complete switch-mode supply in IC form is available.*

of an unregulated DC supply to generate square pulses. These pulses are controlled so that the pulse width (duration) depends on the output voltage required, and the pulses are then rectified and the output smoothed. The frequency of the pulses is high, so that smoothing can be carried out by comparatively low value plastic dielectric capacitors. In the days of discrete circuits, the provision of switch-mode power was too expensive for most purposes, but its advantages of excellent regulation with very low dissipation led to the principle being used for such purposes as TV receiver and computer supplies. As used in IC form, such as the L296, the IC switch-mode regulator allows a very large range of voltage output control, typically 5.1–40 V with inputs of 9–46 V. The switching frequency of, typically, 200 kHz, allows the smoothing components, connected externally, to be of small size.

Other devices

The range of other linear IC devices is very large, and virtually every circuit function that would formerly have been carried out by discrete circuits is now available in IC form. In addition, circuit actions that would formerly have been considered uneconomic to provide in discrete form are available. This has led not only to increased use of linear ICs, but also to increased use of digital ICs, because the effect of ICs on digital circuits has been much greater than that on linear circuits. One important class of ICs falls between the two types inasmuch as it provides for conversion between linear and digital signals. The **A-D converter** will convert linear signals to digital by providing a digital output that is proportional to the voltage amplitude of the input signal. The frequency of the digital signals is determined by a master (clock) oscillator which is usually built into the chip; as an alternative, some chips allow synchronisation to an external oscillator. The precision of the conversion depends on how many bits of digital signal are used, and one common figure is 12-bit, allowing the amplitude of the input signal to be expressed as a number between 0 and 4095. This amount of precision is not always required, and very often a lower resolution is used in order to obtain faster conversion rates so that signals of higher frequency can be converted. Conversion times can be as long as 1.25 ms, which would confine the use of the converter to low frequencies, or they can be as fast as 16 μs. For the conversion in the opposite direction, **D-A converter** ICs can be obtained whose conversion precision is excellent, and which will handle signals of, typically, up to 12 Mhz bandwidth.

Another application of ICs that hovers on the border between linear and digital is the **chopper amplifier**. The principle is illustrated in Figure 13.11, and is used to attain very high-gain DC amplification. A tiny DC signal is converted into a square wave by 'chopping' the signal on and off, using FET switches, and using this square wave as the input to a pulse amplifier. Very high gain of such a waveform can be obtained, and the output square wave can be rectified and smoothed to obtain a high-level DC signal whose amplitude will be very precisely proportional to the amplitude of the original signal.

Finally, one very large application for linear ICs or ICs that make use of both linear and digital techniques, is **instrumentation**. ICs exist to provide for measurement of all the common quantities such as DC voltage, signal peak and RMS amplitude, frequency, etc. Most of the modern ICs of this type provide for digital representation, so that the outputs are in a form suitable for feeding to digital displays.

The most common IC of this type is the **digital DC voltmeter** IC, and a brief account of its action provides some idea of the operating principles of many instrumentation ICs. Referring to Figure 13.12, the voltmeter contains a precision oscillator that provides a master pulse frequency. The pulses from this oscillator are controlled by a gate circuit, and can be connected to a counter. At the same time, the pulses are passed to an integrator circuit that will provide a steadily rising voltage from the pulses.

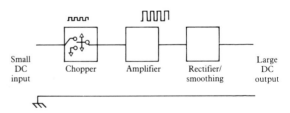

Figure 13.11 *Principle of a chopper amplifier. An oscillator drives a switch that connects the input of an AC amplifier alternately between earth and a small DC input. The amplified square wave is then rectified to provide a DC output, which is proportional in amplitude to the small DC input*

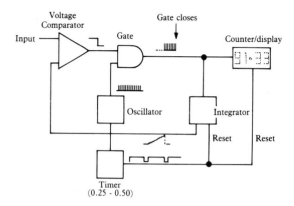

Figure 13.12 *Principle of a digital voltmeter. The oscillator is crystal controlled and its output pulses are passed by the gate to a counter until the output from the integrator equals the input voltage that is being measured. The master timer resets the integrator and the counter so as to repeat the measurement several times per second*

When this voltage matches the input voltage exactly, the gate circuit is closed, and the count on the display represents the voltage level. For example, if the clock frequency were 1 kHz, then 1000 pulses could be used to represent 1 V and the resolution of the meter would be 1 part in 1000, though it would take one second to read one volt. The ICs that are obtainable for digital voltmeters employ much faster clock rates, and repeat the measuring action several times per second, so that changing voltages can be measured. Complete meter modules can now be bought in IC form.

14
Digital integrated circuits

A digital signal is one in which a change of voltage, and the time at which it occurs, are of more importance than the precise size of the change. All of the waveforms in digital circuits are steep-sided, pulses or square waves, and it is the change that is significant. For that reason, the voltages of digital signals are not referred to directly, only as 1 and 0. The important feature of a digital signal is that each change is between just two voltage levels, typically 0 V and +5 V, and that these levels need not be precise. By using 0 and 1 in place of the actual voltages, we make it clear that digital electronics is about numbers, not waveforms.

The importance of using just two digits, 0 and 1, is that this is ideally suited to electronic devices. A transistor, bipolar or FET, can be fully on or fully off, and these two states can be ensured easily, much more easily than any other states. By using just these two states, then, we can avoid the kind of errors that would arise if we tried to make a transistor operate with, say, ten levels of voltage between two voltage levels. By using only two levels, the possibility of mistakes is made very much less. The only snag is that any counting that we do has to be in terms of only two digits, 0 and 1. For some applications, this is of no importance because counting may not be involved. If you use a digital circuit, for example, to control two valves, then the outputs of the digital circuit will turn each valve either fully on or fully off. The action is one of control only, not of counting. On the other hand, we might want to turn one valve on after two pulses at an input, and the other valve on after four pulses at the same input, and this action is quite definitely one of counting.

Digital integrated circuits

Counting with only two digits means using a scale of two. There is nothing particularly difficult about this, because numbers in this scale, called the binary scale, are written in the same way as ordinary numbers (denary numbers or scale-of-ten numbers). As with denary numbers, the position of a digit in a number is important. For example, the denary number 362 means three hundreds, six tens and two units. The positions represent powers of 10, with the right hand position (or least-significant position) for units, the next for tens, the next for hundreds (ten squared), then thousands (ten cubed) and so on. For a scale of two, the same scheme is followed. In this case, however, the positions are for units, twos, fours (two squared), eights (two cubed) and so on. The table of Figure 14.1 shows powers of two and how a binary number can be converted into denary form. Figure 14.2 shows the conversion in the opposite direction.

Digital circuits are switching circuits, and they provide fast switching between the two possible voltage levels. Most digital circuits would require a huge number of transistors to construct in discrete form, so that digital circuits incorporate ICs almost exclusively. These ICs can make use of either bipolar transistors

Power	Number	Power	Number
0	1	9	512
1	2	10	1024
2	4	11	2048
3	8	12	4096
4	16	13	8192
5	32	14	16384
6	64	15	32768
7	128	16	65536
8	256		

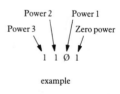

example

Converting from binary into denary.
1 Write down the binary number.
2 For each 1 in the binary number, write down the value for that power from the table above.
3 Add the numbers you have written down.

Figure 14.1 *Binary number places and powers of two, showing how a binary number can be converted to denary*

		Example	
1	Write down the number.	97	
2	Divide it by 2 and write the remainder, which must be 0 or 1, at the side.	48	1
3	Do the same with what is left, the quotient.	24	0
4	Repeat this until only a 0 remains.	12	0
		6	0
		3	0
		1	1
		0	1
5	Now write the remainder digits in order – from the bottom up.	1100001	
6	This is the binary number.		

Figure 14.2 *Converting a denary number into binary*

in integrated form, or MOSFETs, and both types are extensively used. MOSFET types are widely used in computing as memory circuits and in the form of microprocessors. The bipolar types are used where faster operating speeds and larger currents are required, and these applications occur both in computer circuits and in industrial controllers. In this chapter, we shall be looking mainly at the bipolar digital circuits that have been available for a long time and which are still extensively used.

Gates and truth tables

Digital circuits come in several varieties, of which a very important one is the **gate**. A gate in digital electronics is a circuit whose output is a 1 for some specified *combination* of inputs – this type of circuit is sometimes referred to as a 'combinational circuit'. More than 100 years before digital electronics was used, a mathematician called George Boole proved that all of the statements in human logic could be expressed by combinations of three rules which he called OR, AND and NOT. The importance of this is that if we can provide gate actions corresponding to these three rules, we can construct a circuit that will give a 1 output for any set of logical rules. For example, if we want to have an electric motor switched on when a cover is down, a switch is up and a timer has reached zero, or when an override switch is pressed, then this set of rules can be expressed in terms of AND, OR and NOT, and a set of gates can carry out the action.

Digital integrated circuits 223

The action of any gate can be expressed in **truth tables**. The truth table is just a table that shows all the possible inputs to the gate, and the output for each set of inputs. Remember that each input can be 0 or 1 only, so that each input contributes two possible outputs. The total number of outputs is equal to 2^N where N is the number of inputs. For example, if there are four inputs to a gate, then the number of possibilities is $2^4 = 16$, and a truth table will consist of 16 lines. For a lot of truth tables, there is only one output that is different from the rest, and it is easier to remember which one this is than to try to remember the whole of a truth table.

Figure 14.3 shows truth tables for the basic AND, OR and NOT gates. Of these, the NOT gate is a simple one, with just one input and one output. Its action is that of an inverter. If the input is 0, then the output is not-0, which is 1. If the input is 1, then the output is not-1, which is 0. The other two permit more than one input, and the examples show two inputs, the most common number. The action of the AND gate is to give a 1 output when

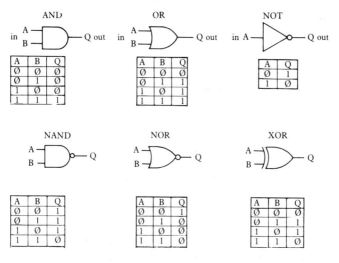

Figure 14.3 *Truth tables and symbols for AND, OR and NOT gates. These are the basic logic gates, but gate ICs concentrate mainly on inverting types, such as NAND and NOR, along with the XOR gate whose action is slightly different from that of the OR gate. The value of the NAND or NOR gate is that either one of these gate types can be used in combination to provide any form of gate action*

both inputs are at 1, and a 0 output for any other combination. The action of the OR gate is to give a 0 output when each input is 0, but a 1 for any other combination of inputs. The same arguments apply to gates with more than two inputs, and the illustration also shows the truth tables for the common NAND, NOR and XOR gates.

Sequential circuits

The other basic type of digital circuit is the **sequential**. The output of a sequential circuit, which may mean several different output signals on separate terminals, will depend on the *sequence* of inputs. The simplest example is a counter IC, in which the state of the outputs will depend on how many pulses have arrived at the input. The basis of all sequential circuits is the **flip-flop**, and though simpler types exist, the most important type of flip-flop is that known as the master-slave J-K flip-flop, abbreviated mercifully to JK. This is a clocked circuit, as the action of the IC is carried out only when a pulse is applied to an input labelled 'clock'. This allows the actions of a number of such circuits to be perfectly synchronised, and avoids the kinds of problems that can arise in some types of gate circuits when pulses arrive at different times. These problems are called 'race hazards', and their effect can be to cause erratic behaviour when a circuit is operated at high speeds. When clocking is used, the circuits are synchronous, meaning that each change takes place at the time of the clock pulse, and there should be no race hazards.

The symbol and a **state table** for a JK flip-flop is illustrated in Figure 14.4. The table is a state table rather than a truth table, because the entries show that it is possible to have conditions in which identical inputs do not produce identical outputs, since the outputs depend on the sequence of inputs rather than the voltages of the inputs at any particular time. The inputs are to the J and K terminals, and the outputs are taken from the Q and Q̄ terminals. The Q̄ output is always the inverse of the Q output. The table shows the possible states of inputs before and after the clock pulse, showing how the voltages on the J and K pins will determine what the outputs will be after each clock pulse. Note that if both the J and K terminals are kept at level 1, the flip-flop will toggle so that the output changes over at each clock pulse, as is required for a simple binary counter.

Digital integrated circuits 225

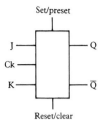

$\bar{Q}$ output is always inverse of Q output
Set/preset input makes Q = 1 at any time
Reset/clear makes Q = ∅ at any time
Other inputs at J and K have effect only when clock
terminal Ck is pulsed.

Type of change	J	K	Q_n	Q_{n+1}
no change	∅	∅	∅	∅
			1	1
reset	∅	1	∅	∅
			1	∅
set	1	∅	∅	1
			1	1
toggle	1	1	∅	1
			1	∅

Q_n is output at Q just before clock pulse
Q_{n+1} is output at Q just after clock pulse

Figure 14.4 *The symbol for a J-K flip-flop and its state table. The set/reset pins are asynchronous, meaning that their action is independent of any clock pulse. The inputs at the J and K terminals have effect only when a clock pulse is applied to the Ck terminal*

Flip-flops are the basis of all counter circuits, because the toggling flip-flop is a single stage scale-of-two counter, giving a complete pulse at an output for each two pulses in at the clock terminal. By connecting another identical toggling flip-flop so that the output of the first flip-flop is used as the clock pulse of the second, a two-stage counter is created, so that the voltages at the Q outputs follow the binary count from 0 to 3, as Figure 14.5 shows. This principle can be extended to as many stages as is needed, and extended counters of this type can be used as timers, counting down a clock pulse which can be at a high frequency initially. The type of counter which uses toggling flip-flops in this way is **asynchronous**, because the last flip-flop in a chain like this cannot be clocked until each other flip-flop in the chain has changed. For some purposes, this is acceptable, but for many other purposes it is essential to avoid these delays by using synchronous counter circuits in which the same input clock pulse is applied to all of the flip-flops in the chain, and correct counting

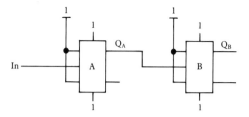

Input pulse no.	Q_A	Q_B	Binary no.
0	0	0	00
1	1	0	01
2	0	1	10
3	1	1	11
4	0	0	00

Figure 14.5 *Using the flip-flops in a counter circuit. The outputs of the flip-flops provide the digits of a binary number which represents the number of pulses at the first input. One flip-flop is needed for each digit in a counter. Note that circuit conventions require the pulse input to be shown entering at the left hand side, but the binary number is written the other way round*

is assured by connecting the J and K terminals through gates. Circuits of this type are beyond the scope of this book, and details can be found in any good text of digital circuit techniques.

The 74 series of TTL ICs

The letters TTL stand for **transistor-transistor logic**, as originally applied to a form of bipolar logic circuit for gates and flip-flops, and members of this family of devices have always been distinguished by the use of numbers and letters which start with '74'. The original TTL circuits used bipolar transistors in arrangements in which the logic signal input was always taken to the emitter of a transistor whose base was connected to supply positive (Figure 14.6). This had the effect that no current flowed at the input for a logic 1 signal, but a current of about 1.6 mA flowed out from the input when the input signal was at level 0. The original types of TTL ICs are still available, but have been rendered obsolete by a range of devices with identical actions but better specifications. To see what features of IC gates and flip-

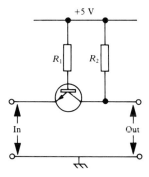

Figure 14.6 *The type of input stage used for the original TTL circuits. This is a common-base circuit, so that no current flows at the input when the input is at logic 1, but about 1.6 mA (determined by the integrated resistor R_1) will flow from the input when the input is held at logic 0*

flops are important, however, we can use the figures for the old standard TTL types.

Two important features of any IC for digital purposes are the propagation time and the power consumption per device. The **propagation time** is taken as an average figure for a gate, and is the time between changing the inputs and getting a change at the output. The original types of TTL featured propagation delays of the order of 10 ns. This is acceptable for all but the fastest circuits, and so newer forms of TTL circuits have aimed at about this propagation time. To achieve it the standard TTL circuits used power dissipation figures in the region of 10 mW per gate. This sounds low, but for a device with several hundred gates of gate-equivalent circuits, the power consumption could be difficult to dissipate, and resulted in the chip running hot. The problem with the original type of design was that power dissipation and speed were connected – any reduction in one could be achieved only at the expense of the other.

The other feature of standard TTL circuits was the amount of current required to drive an input to logic 0. The original TTL ICs required 1.6 mA to be passed from an input at logic zero, and the device which had to drive the TTL circuit had to be capable of **sinking** this current, i.e. keeping the input voltage at logic level 0 while this amount of current flowed into its output stage. This made it much better to drive a TTL input from the collector of a transistor rather than from an emitter-follower (Figure 14.7)

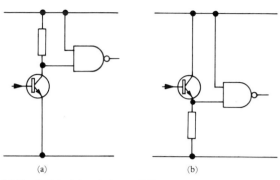

Figure 14.7 *Driving signals into a TTL gate. The common-emitter stage (a) is preferable if the gate is an original TTL type because an emitter-follower (b) places a resistance between the TTL input and earth*

because the collector circuit could pass this current and still remain at about 0.2 V, while the resistor of the emitter follower would have a voltage across it due to the current; a 470R resistor would, for example, have a voltage of 0.75 V across it with 1.6 mA flowing, and this voltage is much too close to the absolute upper limit of 0.8 V for logic level 0.

Conditions are easier when one logic circuit provides the output for another, because standard TTL was designed to be able to sink up to 16 mA at its output. This amount was guaranteed, and many types were also able to supply, or **source** the same amount of current out from the output terminals, although this was not guaranteed. By being able to sink 16 mA, each TTL output could be connected to 10 TTL inputs and could be guaranteed to drive all of them to logic 0. This figure of 10 is called the **fan-out** for an output. There is a less important figure, the **fan-in**, meaning the load current, compared to standard TTL, at the input for logic 0.

The conditions of use for standard TTL circuits are quite stringent, particularly the supply voltage. This has to be maintained at 5.0 V, with a tolerance of about 0.25 V and an absolute maximum voltage of 7.0 V. Each time a gate changes over, a voltage spike will appear at the supply terminals because there is a very brief short-circuit during switch-over. This voltage spike can cause problems in other TTL circuits, and so the 5.0 V supply should always be stabilised. In addition, it is good practice to connect a capacitor of about 10 nF between the supply and earth pins of each device, though if low-power devices are being used this can be relaxed to one capacitor for each five devices. The capacitors are particularly important in large circuits in which there may be long paths from the voltage regulator IC to the TTL circuits.

The propagation times for TTL circuits are affected by temperature, and by the types of load that can be connected. The worst type of load is a capacitive load with a pull-up resistor, as illustrated in Figure 14.8, because to reduce the voltage of this output to logic 0 means passing enough current through the resistor and also discharging the capacitor, even if the capacitance is only stray capacitance. The switching speed of the circuit will be affected by the amount of capacitance, typically by about 7 ns for 80 pF. This switching speed may also be affected by changes of temperature, but this effect is comparatively small.

The noise immunity of a digital circuit is an important feature because it is a guide to what amplitude of interference may cause erratic operation of the circuits. Noise immunity is often quoted in terms of the size of interference pulse that will just cause a circuit to switch-over. For TTL circuits, the noise immunity is determined by the small difference between 0 and 1 levels. If, for example, the maximum logic 0 voltage of 0.8 V is used, and if the gate actually switches over at 1.4 V (a fairly common figure), then an unwanted pulse of 0.6 V will be enough to cause trouble. This makes TTL circuits very susceptible to interference on the supply lines, and is another good reason for using capacitors across the supply and earth terminals of each device.

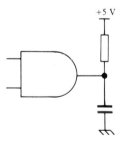

Figure 14.8 *The worst type of TTL output load: a resistor to +5 V and a capacitor to earth. This is generally the type of load that will be connected if the output of a gate drives another gate input*

Schottky TTL

Standard TTL devices had been in use for some time when a considerable advance was made in the form of **Schottky TTL**. The name arises from the device called the **Schottky diode**, a diode made using semiconductor and metal (usually aluminium) which has a very low forward voltage when conducting. One of the problems about standard TTL is that a considerable amount of power is wasted in holding transistors fully on. In addition, a bipolar transistor that is switched fully on, with its collector voltage as low as it can be taken, cannot be rapidly switched off. This is because the base region of the transistor is full of carriers of the minority type (electrons in the base region of the *n-p-n* transistor, for example), and these take some time to clear when the bias is reversed. It is because of this 'saturation' that any attempt to obtain faster switching of bipolar transistors always results in high power dissipation. Very much lower power dissipation and faster switching can be obtained if each transistor is prevented from becoming saturated. This is achieved by using Schottky diodes in the gate circuits, arranged so that if the voltage at the base of a transistor becomes too high, current is diverted to the collector so as to prevent saturation. In addition, these diodes can be used in the inputs of gates in preference to the emitter leads of transistors, and this allows much lower currents to be used, so that much less current has to be sunk to obtain a level 0 input. The original TTL LS (low-power Schottky) devices required only about 0.4 mA to be sunk at each input, and gates could have a fan-out of 10–20. Power consumption dropped from the 10 mW per gate level to about 2 mW per gate, an improvement of five times, but with virtually the same propagation times maintained.

There is a very large family of TTL LS circuits available now, more in the LS variety than in the original type of TTL. At the same time, however, a huge range of other methods of forming gates has been developed, resulting in a bewildering range of devices. None of these options offers the full range of gate and flip-flop types, but most ranges contain all of the popular devices. A few devices that carry 74 series numberings are not bipolar but field effect devices, and are noted in the following section. Of the true Schottky TTL types, the more notable are the AS, F and

ALS. The AS (advanced Schottky) and F (fast) types are designed for faster switching, in the region of 4 ns, with a power consumption in the region of 6 mW per gate. This power dissipation is less than that of the standard TTL, but with significantly faster switching. The ALS type is a different compromise, with a power level of about 1 mW per gate, but a switching time of around 8 ns, marginally faster than the LS types. Figure 14.9 shows a list of the 74 series types that are available in the different forms from RS Components Ltd.

CMOS ICs

The acronym CMOS stands for **complementary MOS**, complementary in this sense meaning that MOS transistors with both types of channels are used in pairs. A typical MOS circuit will use a p-channel FET in series with an n-channel FET, as illustrated in Figure 14.10. In this simplified basic circuit, the gate terminals of the FETs are connected together, so that an input signal, which will be a digital signal, will be applied to both gates. If we imagine that the input signal is at logic 1, then the result will be to make FET A conduct fully, and FET B cut off, with the result that the output is connected to the supply line at voltage level 1. If the input to the gates is at level 0, then the FET A is cut off, and the FET B is fully conducting, connecting the output to zero voltage, logic level 0. The basic circuit illustrated here is therefore one whose output is a replica of its input, neither inverting nor carrying out any other type of gating action. A circuit of this type can be used as an **expander**, allowing a larger number of loads to be driven from a single output.

The advantages of CMOS construction are that a wide range of supply voltages and logic voltages can be used. The standard and LS TTL circuits are designed to be operated with a +5 V supply, and this voltage must be adhered to very closely. Most CMOS circuits will operate with voltages as low as 3 V, some at 1.5 V. They can also be operated with voltage levels of 12 V or more, so that CMOS circuits can be used with low-voltage battery power equipment, and also can be applied with the voltage levels that are encountered in a car. The switching is also to levels that are very close to the supply voltages. The construction of the output stages of TTL circuits makes it difficult to achieve a voltage for logic

	Type	LOGIC FAMILY					
		STD	LS	ALS	HCT	HC	FAST
gates							
AND							
Quad 2-input	08	*	*		*	*	*
Quad 2-input o.c.	09		*	*			
Triple 3-input	11		*	*	*	*	*
Triple 3-input o.c.	15		*				
Dual 4-input	21		*	*			
OR							
Quad 2-input	32	*	*	*		*	*
Exclusive OR							
Quad 2-input	86	*	*		*	*	*
Triple 3-input	4075					*	
NAND							
Quad 2-input	00	*	*	*	*	*	*
Quad 2-input o.c.	01	*					
Quad 2-input o.c.	03	*	*	*		*	
Triple 3-input	10	*	*	*	*	*	*
Dual 4-input	20	*	*	*		*	*
Dual 4-input o.c.	22		*				
Quad 2-input high voltage	26		*				
8-input	30	*	*	*		*	
Quad 2-input buffer	37	*	*	*			*
Quad 2-input o.c.	38	*	*	*			*
Dual 4-input buffer	40	*	*				*
13-input	133		*			*	
NOR							
Quad 2-input	02	*	*	*		*	*
Dual 4-input with strobe	25	*					
Triple 3-input	27	*	*	*	*	*	*
Quad 2-input buffer	28	*		*			
Quad 2-input buffer	33		*	*			
Quad 2-input exclusive	266					*	
Dual 4-input	4002					*	
8-input	4078					*	
Schmitt							
Dual 4-input NAND	13	*	*				*
Hex inverting	14	*	*			*	*
Quad 2-input NAND	132	*	*			*	*
AND-OR-Invert							
Dual 2-wide, 2-input	51	*	*				*
4-wide	54		*				
2-wide, 4-input	55		*				
4-2-3-2-input	64						*
Buffers							
Hex inverting	04	*	*	*	*	*	*
Hex inverting o.c.	05		*	*			
Hex inverting o.c.	06	*					
Hex o.c.	07	*					
Hex inverting o.c.	16	*					
Quad 3-state active low enable	125	*	*			*	*
Quad 3-state active high enable	126	*	*			*	*
Hex 2-input NOR enable	365	*	*			*	*
Hex inverting, 2-input NOR enable	366		*			*	*
Hex 3-state	367	*	*			*	*
Hex 3-state inverting	368		*			*	*
line/bus, drivers/ transceivers							
Quad line driver	128	*					
Octal buffer 3-state inverting	240		*			*	*

Digital integrated circuits 233

	Type	LOGIC FAMILY					
		STD	LS	ALS	HCT	HC	FAST
Octal buffer 3-state	241		★		★	★	★
Quad bus transceiver inverting	242		★			★	
Quad bus transceiver	243		★			★	
Octal buffer 3-state	244		★		★	★	★
Octal bus transceiver 3-state	245		★			★	★
Quad tridirectional transceiver true	442		★				
Quad tridirectional transceiver inverting	443		★				
Quad tridirectional transceiver	444		★				
Octal bus transceiver	620		★				
Octal bus transceiver	640				★	★	
Octal bus transceiver	643				★	★	
Octal buffer with parity	655A						★
Octal buffer with parity	656A						★
Octal transceiver with parity	657						★
Quad bus transceiver	1242						★
Quad bus transceiver	1243						★
Hex inverting buffer	4049					★	
Hex buffer	4050					★	
flip flop (bistables)							
D-type							
Dual edge triggered	74	★	★	★	★	★	★
4-bit	75	★	★			★	
4-bit registers	173					★	
Hex with clear	174	★	★		★	★	★
Quad with clear	175	★	★			★	★
Octal 3-state	374		★			★	★
Octal common enable	377				★		
Hex	378		★				★
Octal inverting latches	533					★	
Octal inverting latches	534					★	
Octal transparent latch	573			★	★	★	
Octal edge triggered	574					★	
Octal transparent latch inverted	580			★			
J–K							
AND gated positive edge triggered	70	★					
AND grated master slave	72	★					
Dual with clear	73	★	★			★	
Dual with preset and clear	76	★	★			★	
Dual with present common clear and clock	78		★				
Dual with clear	107	★	★		★	★	
Dual positive edge triggered preset and clear	109		★				★
Dual negative edge triggered preset and clear	112		★	★		★	
Dual negative edge triggered preset	113		★			★	
Dual negative edge triggered preset and clear	114		★				
Monostable multivibrators							
Single	121	★					
Dual retriggerable with clear	123	★	★			★	
Dual	221	★					
Dual retriggerable	423					★	
Dual	4538					★	

	Type	LOGIC FAMILY					
		STD	LS	ALS	HCT	HC	FAST
Latches							
Dual 4-bit addressable	256		*				
8-bit addressable	259	*	*			*	*
8-bit register with clear	273	*	*			*	*
Quad 2-port register	298		*				
Octal 3-state	373		*			*	*
Quad 2-port register	399						*
Octal transparent inverted outputs	563					*	
Octal edge triggered inverted outputs	564					*	
arithmetic functions							
4-bit full adder with carry	83A		*				
4-bit magnitude comparator	85		*			*	*
4-bit arithmetic logic unit	181		*				*
4-bit full adder with carry	283	*	*				*
4-bit arithmetic logic unit	381						*
4×4 register file 3-state	670		*				
8-bit magnitude comparator	682		*				
Magnitude comparator 8-bit	688					*	
counters							
Decade up	90	*	*				
Divide by 12	92		*				
4-bit binary	93	*	*				
B.C.D. asynchronous reset	160		*			*	
Binary asynchronous reset	161	*	*			*	*
B.C.D. synchronous reset	162		*			*	
Binary synchronous reset	163		*			*	*
4-stage synchronous bidirectional	169						*
Binary up/down synchronous	191	*	*				
Decade up/down synchronous	192	*	*			*	
Binary up/down synchronous with clear	193	*	*			*	
Decade presettable ripple	196		*				
4-bit presettable ripple	197		*				
Dual decade	390		*			*	
Dual 4-bit binary	393	*	*		*	*	
4-bit binary	669		*				
Decade/divider	4017					*	
14-bit binary	4020					*	
12-bit binary	4040					*	
14-bit binary	4060					*	
shift registers							
4-bit	95	*	*				
5-bit	96	*					
8-bit serial in parallel out	164		*		*	*	*
8-bit parallel to serial	165		*		*	*	
4-bit universal	194		*			*	*
4-bit parallel access	195	*				*	
8-bit universal storage 3-state	299		*			*	
8-bit universal storage 3-state	323		*				
4-bit cascadable	395		*				*
16-bit serial to parallel	673		*				
16-bit parallel to serial	674		*				

Digital integrated circuits   235

	Type	LOGIC FAMILY					
		STD	LS	ALS	HCT	HC	FAST
encoders, decoders/ drivers							
Decoders							
B.C.D. – decimal	42	*	*		•	*	
B.C.D. – decimal driver	45	*					
B.C.D. – 7-segment driver o.c.	47	*	*				
B.C.D. – 7-segment driver	48		*				
B.C.D. – 7-segment driver o.c.	49		*				
De-multiplexer	137		*				
3 to 8 line multiplexer	138		*		*	*	*
Dual 2 to 4 line multiplexer	139		*		*	*	*
B.C.D. – decimal driver	141	*					
B.C.D. – decimal driver	145		*				
4 to 16 line	154	*			*	*	
Dual 1 of 4	155	*	*				
Dual 1 of 4 o.c.	156		*				
BCD to 7-segment latch/driver	4511					*	
4-bit latch 4 to 16 line	4514					*	
BCD 7-segment latched LCD driver	4543					*	
Encoders/multiplexers							
Priority encoder 10-line decimal to 4 line BCD	147					*	
Octal priority encoder 8 to 3	148						
8-input multiplexer	151	*	*		*	*	*
Dual 4-input multiplexer	153	*	*			*	*
Quad 2-input multiplexer	157		*			*	*
Quad 2-input multiplexer inverting	158	*				*	*
Parity generator/checker 9-bit odd/even	180		*				
Selector multiplexer 3-state	251	*	*		*	*	
Dual 4-input multiplexer 3-state	253		*			*	*
Quad 2-input multiplexer 3-state	257		*			*	*
Quad 2-input multiplexer 3-state inverting	258		*				
9-bit parity generator/ checker	280					*	*
Dual 4-input multiplexer inverting	352		*				
Dual 4-input multiplexer 3-state inverting	353		*				
Data selector multiplexer transparent	354		*			*	
Data selector multiplexer	356		*			*	
miscellaneous							
Crystal oscillator	321		*				
Voltage controlled oscillator	625		*				

o.c. – open collector input

Figure 14.9 *The variety of 74 series logic ICs available in different forms, courtesy of RS Components Ltd*

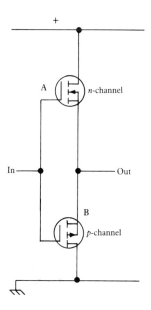

Figure 14.10 *A typical CMOS gate stage using both n-channel and p-channel FET construction. If the positions of the devices are interchanged, the gate becomes an inverter*

level 1 that is close to the +5 V supply level, and which is more likely to be typically around 3.5 V. This makes for a rather poor noise immunity for TTL, and because CMOS can achieve logic levels which are almost equal to the supply voltages, the noise immunity of CMOS can be very much better than that of standard or LS TTL. In addition, the power consumption of CMOS can be very much lower than that of TTL, of the order of a microwatt per gate.

The disadvantages of CMOS are the slow operating speed of some types, though this is not totally inevitable, and the need for static protection. The typical propagation time for the normal type of CMOS gate is between 100 and 200 ns, ten times as much as the average time for the TTL type of gate. For many applications, this is of no importance at all; the time that is needed for a pocket calculator to work out an action is seldom found to be too long. Where the speed of CMOS may be found unacceptable is in large circuits in which many actions have to be performed in sequence, so that the slow speed of CMOS will greatly increase the overall time as compared to TTL. The gates and flip-flops of the 4000 series of CMOS circuits are typical of the older type of CMOS technology in this respect. Modern CMOS devices can, however, be manufactured with very much faster operating speeds, and we shall look at some of these shortly. The need for faster CMOS has been made necessary by the need for some memory chips in computers that can use very low power, and retain data when operated from a low-voltage battery.

In the early days of CMOS, electrostatic damage caused a lot of problems. The gates of MOS devices are so well insulated from the channels that even a tiny amount of electrostatic charge would not be discharged quickly, so permitting a build-up of voltage. At the same time, the very thin insulating layer of silicon oxide was easily punctured by excessive voltages, of the order of 20 V or less in some cases. This means that handling a MOS device, or rubbing the pins against any insulating material or against textiles could cause total and irreversible breakdown of the gates. By the time that CMOS ICs were being used in any numbers, the circuits included discharge diodes at the gate terminals that ensured voltage limitation. The important point to remember, though it is never mentioned by the manufacturers of 'anti-static' aids, is that only voltage between the gate and the channel can cause damage. It is possible that by walking across a nylon carpet, your hands can be at 15 kV or more, but if this voltage is applied to both gate and channel, it does no harm. ICs that are correctly connected into a circuit are therefore immune from these problems, because the circuits will have some resistance connected between gate and channel connections that will safely limit any conceivable electrostatic voltage. In other words, static should not cause any damage to a working circuit, whether switched on or off. The only possible case in which damage can occur in normal use is when one pin of a MOS IC is earthed and another pin touched, and even in this case, the built-in diodes should be able to cope. In many years of handling MOS devices, I have never had one damaged by static, even on days when walking across the carpet and touching a radiator would result in a numbing shock.

Figure 14.11 lists the standard precautions for handling MOS devices. These err, as always, on the safe side, and the precautions that are used in some assembly lines (earthed metal

> *Precautions for using MOS devices*
> 1. Keep all MOS devices in the manufacturer's packing until needed. Try to avoid touching leads/pins.
> 2. Short pins or leads together while installing if possible.
> 3. Never allow a gate input to become open-circuit.
> 4. Use earthed soldering irons, and earth circuit boards while installing devices.
> 5. Ensure that circuit boards contain resistive paths between gate and source or drain terminals.
> 6. If devices have to be handled, ensure that surroundings are suitable – high humidity, no carpets of synthetic fibres, earthing on all metal surfaces.

Figure 14.11 *Standard handling precautions for FET and MOS devices generally. All devices of this type are most vulnerable to damage when there is no conducting path between the gate and the source of any FET, but ICs are generally protected by built-in overload diodes*

manacles for anyone handling MOS chips, for example) are not mentioned because they apply to specialised situations only. If ICs are never inserted or removed until equipment has been switched off and all voltages discharged, if ICs are kept in their protective packaging when not in use, and if minor precautions are taken against static, like working with the hands slightly damp and with all circuit boards earthed, then problems due to static are most unlikely.

Figure 14.12 shows the general form of a CMOS gate. The input is always to gate terminals, and the output is taken from one drain and one source. Because the gate terminal of a MOS IC does not require current, the fan-out of CMOS gates is limited only by the capacitance of the gate terminals. In other words, if the operating speed is very low, the fan-out is almost unlimited, but for the higher operating speeds, the amount of current that has to be supplied in order to charge and discharge the gate-channel capacitance will be a limiting factor. The presence of the protective diodes also increases the amount of DC current leakage, though by a very small amount.

The HC and HCT series

The older 4000 series of CMOS circuits were very useful, but their low operating speed proved to be a handicap in some types of circuits. This led to the development of faster circuits which were

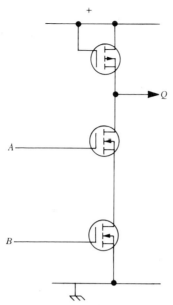

Figure 14.12 *A typical CMOS gate, in this case a NAND gate. When the A or B input is low, there is no conducting path between Q and earth, but the load FET will conduct sufficiently to keep the voltage at Q high. When both of the n-channel FETs conduct, the voltage at Q will drop, because the p-channel FET will be pinched off*

designed from the start to be pin-compatible with the well-established TTL LS families. The first types to be developed were the 74HC family, which offered operating speeds identical with, or slightly faster than, LS TTL, but with power consumption that was one thousand times lower. This makes the types less compatible electrically – one HC gate output can drive only one LS input. The 74HC series use the same type numbers as their other TTL counterparts, and can be operated with supply voltage level of +2 V to +6 V. The later 74HCT series are designed for close compatibility, and are designed to operate with a supply of +5 V, with a 10 per cent tolerance. The number of devices that can be bought in the HCT series is much more limited, however.

Though CMOS forms one important class of digital ICs, notably for its very low power consumption, many digital ICs use MOS FETs of one channel type only, such as *n*-MOS and *p*-MOS. These ICs offer low power consumption, though not as low as that of CMOS, along with fast operation. These types of FET

ICs are used extensively in computing, both for microprocessors and for the supporting memory chips. These and CMOS ICs are also used in a variety of other applications for instrumentation and communications. Many of these ICs are of familiar types, such as counters, but with a greater degree of integration, such as the 50395 6-decade counter. Some are of a very specialised nature, like the speech chips that can be incorporated into equipment to permit spoken messages to be delivered. Many of the p-MOS and n-MOS ICs are for computing, and the popular microprocessor types are included in this heading.

As well as the use of microprocessors and memories, however, dealt with later in this section, computing requires interface chips such as drivers for floppy-disc units, and the keyboard and VDU interface chips. Keyboard interface ICs will have on-board circuits that suppress the effect of the contact bounce that plagues mechanical switches, but for other applications, separate bounce eliminator ICs are available. In this category of assorted circuits, we also have the 'bucket brigade', which is a type of register in which data is shifted from one device to another in a chain at each input (clock) pulse. This device allows information to be delayed by a time that depends on the frequency of the clocking pulses, and the main uses are in obtaining delays for audio systems such as public address, music reverberation and special effects.

An important feature of many of the more advanced MOS devices is three-state operation. This does not imply that there is a third state of data in addition to the 0 and 1 levels, only that the devices can be isolated with outputs floating. The third-state switching is achieved by a single pin which allows the device to be fully active, or put into a floating state with the output pins disconnected. The use of this floating state allows several devices to have their outputs connected together, sharing the same lines but with only one device controlling the voltage of a line at any given time. This method of operation is particularly important for computing ICs, because the lines are shared between many devices, and are known as **bus lines** (from the Latin, *omnibus*, meaning *for all*).

Memory ICs

One very important class of MOS ICs for computing use is the memory type. Memory in computing terms means the ability to retain logic 0 or 1 signals for as long as is needed, and memory may be **volatile** or **non-volatile**. Volatile memory is any type that requires a power supply to be switched on in order to retain data. All electronic memory is of this type, though some CMOS memory can be arranged to retain data almost indefinitely with the help of a battery that is built into the chip. The true non-volatile memory will retain data with all forms of voltage supply switched off. The most widely used form of non-volatile memory is the magnetic type, and the older type of large (mainframe) computers used tiny rings or cores of magnetic material, each one retaining one binary digit (or **bit**). For storage outside the computer, magnetic tapes and discs are used, and a more recent development is the optical type of disc as used in the CD recording, written and read with a laser. The advantage of the CD disc is that a very much larger amount of data can be stored in a very compact form, and with no risk that the data can be erased by careless exposure to magnetic fields, or by clumsy handling.

Memory within the computer is nowadays mainly volatile memory, however, because it is relatively easy to transfer data to magnetic discs or tapes for longer term storage. The types of volatile memory are classed as **static** or **dynamic**, and the two are radically different. Static memory (usually referred to as static RAM, because access to the memory can be random, to any part of the memory) makes use of a flip-flop for each bit that is to be stored. Since a flip-flop consists of two transistors, one of which is passing current while the other is off, each flip-flop in a memory of this type will pass current whether the bit that is stored is a 1 or a 0. Very early types of memory of this sort used the flip-flops in a chain, clocked so that each clock pulse would shift each bit of data to the next flip-flop in line. This made for a very simple type of memory system consisting of just one flip-flop for each bit, but with limitations to access. For example, 1000 bits of data would be stored in a memory, assuming that it had at least 1000 flip-flops available, by applying the bits to an input along with a clock pulse for each input change. If the data was needed again, it had to be obtained at the output end of the chain, one bit for each clock pulse. A system like this is called first-in first-out, or **FIFO** memory. Each time the data is read out it must also be placed back into the flip-flops again if it is to be held for further use. In addition, all of the data has to be read; it is not possible to read the 56th bit without reading the first 55, for example, and if you need to keep the data then reading and writing must be carried out together.

Developments in IC techniques soon made it possible to connect inputs and outputs to any of the flip-flops in a chip. This is called **random access**, and the name of **random-access memory** (RAM) was applied to such a chip to distinguish it from the earlier types. Random access demands the use of large numbers of gates on the chip, and also some method of selecting which flip-flop is to be accessed. This is achieved by applying binary signals to address pins. The address of a bit is its reference number, the number of the flip-flop in which it is stored. When this number is selected by applying the binary signals for the number to a set of address pins on the chip, the correct gating is selected to make connection to that flip-flop, either for reading or for writing. The basic type of RAM chip therefore consists of one input pin, one output pin, a read/write pin, and a number of address pins. The use of 16 address pins, for example, allows $2^{16} = 65536$ different flip-flops to be accessed, so that this number of bits can be stored. Any one flip-flop can be accessed by using its particular address number in binary form on the address pins.

The use of flip-flops requires current in each flip-flop, and early types of static memory required a considerable amount of power. This led to the development of a different memory type, the dynamic memory, and though CMOS static memory can now be obtained with very low power consumption, the use of dynamic memory is very well established, particularly in small computers. The principle of dynamic memory is storage of charge in a tiny capacitor constructed like the gate of a MOS transistor. Though the resistance between the contacts of such a capacitor can be very high, the small amount of charge that can be stored, along with the leakage through the selecting circuits of a memory chip, limits the storage time to a few milliseconds only. This means that each tiny capacitor that stores a 1 must be recharged (or **refreshed**) at intervals of, usually, one millisecond. This is not as difficult as might be imagined, because the memory chips can be manufactured with the refreshing circuits built-in, and the master control of refreshing can be supplied either from a special controller chip, or in some cases by the microprocessor of the computer itself.

The refresh has to be 'transparent', meaning that it should not interrupt the normal computing actions in any way.

The use of dynamic memory greatly reduces the amount of power that is needed for the memory of a small computer, because power is needed only to refresh the charge on the capacitors that are storing a logic 1 signal. The low power requirement makes it easier to build very large-capacity memory chips, so that the spectacular improvement in the memory that is available for small computers has been due largely to the availability of dynamic memory chips in ever increasing sizes. At the time of writing, the largest dynamic RAM in common use was of 262144 bits. This is described as a 256K chip, because 'K' in computing is used to mean 1024 (which is 2^{10}) bits.

Memory organisation

Computers do not deal with single bits, and the most common groupings are of 8 bits (one byte), 16 bits (one word) or 32 bits (double-word). A computer will therefore need to make access to at least 8 bits of memory at a time, so that memory has to be arranged in circuits that make this possible. If we deal only with 8-bit principles here, it makes the diagrams simpler, but the principles can easily be extended to apply to the 16- and 32-bit machines that are becoming more common. One simple organisation method is to have one memory chip handle one bit in each group. Suppose, for example, that you are using a group of eight, 1 byte, as the unit for the computer, and your memory requirement is 64K (65536 bytes). This can be achieved by using 16 address lines, and if these address lines are connected to each and every memory chip, then we can use memory chips that deal with 64K bits each. By using eight such chips (Figure 14.13), we can obtain access to one bit of the byte in each chip. The data to or from the memory are connected by a data line connected to each chip, and the set of eight lines is called the **data-bus**.

This is by no means the only possible arrangement of memory chips, but the only common alternative is the use of four bits per chip. A chip of this type will use four data lines, and as many address lines as will be needed for the number of groups of four that will be used. For example, the popular 256K chip uses 64K groups of four bits, so that only two chips are needed to form 64K

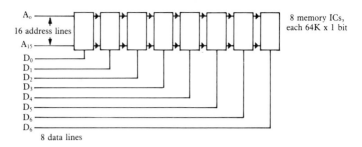

Figure 14.13 *The common arrangement of memory chips for small computers in the past: 8 chips each of which can store 1 bit at each of 64K addresses*

of memory (Figure 14.14). The important point is that memory systems are connected with all of the address lines of the address-bus connected to a set of memory chips, and each chip contributing one or more lines of the data-bus.

For the modern small computers that use more than 64K of memory, various methods of connecting the additional memory may be used. One method is **banking**, in which sets of memory chips are switched over so as to select one of a number of banks of 64K. This is a system that is extensively used when only 16 address lines can be obtained from the microprocessor. When more address lines, typically 24, can be used, the memory will still be arranged in banks of 64K, but with the address lines gated so that different banks are selected by the lower 16 lines of the address bus. Figure 14.15 gives a simple example of the use of 128K in such a system. These switchings are made possible by the use of 'enable' pins on the memory chips that allow chips to be

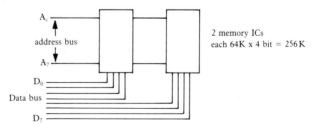

Figure 14.14 *The alternative arrangement using 256K chips of 64K × 14 bit. The 256K chips are extensively used in constructing large memories for the modern 16-bit computers*

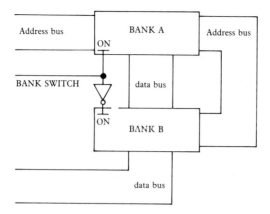

Figure 14.15 *Switching memory banks. The memory chips are arranged in banks, each of 64K, sharing the same address and data lines. The enable pins on the chips are connected in each bank, and these are used to switch from one bank to another. The use of an inverter in this example ensures that both banks can never be activated together*

switched completely on or off by the logic voltage at such pins. A few chips have several enable pins along with internal gating that permits an AND or OR action at the enable pins.

Reading and writing of data is done using the same set of data lines, so that these operations can never be carried out at the same time. Each memory chip will have a read/write pin, sometimes a pair of pins, whose logic voltage or voltages will decide which action is to be carried out. Only RAM memory chips will need these pins, because the RAM principle allows both reading and writing of the memory. All computers will require some memory that is completely non-volatile. This is because the action of the keyboard and the screen depend on a program being present in the memory of the computer. If nothing is held in the memory of the computer at the time it is switched on, then the machine cannot be used, since nothing is retained in the RAM after a switch-off. It is therefore necessary to have some part of the memory non-volatile. In some machines, this part of memory can be very extensive, as much as half of the total memory. More commonly now, the computer is a 'clean machine', meaning that the non-volatile part of memory is very small, only large enough to load in data from a disc. This data can then consist of a program (the **DOS**) that will take care of 'housekeeping' actions like the action of the keyboard, screen and the disc drives.

Memory of this type is referred to as ROM, **read-only memory**, and it can be of several forms. Where the machine uses an operating program that is well-established, like CP/M, MS-DOS, Unix, etc., the ROM that is used for loading the remainder of the program will be similar for a large number of computers. It can therefore be programmed during manufacture, so that the gates that are operated by the address pin voltage will simply connect the data pin(s) to logic level 0 or 1 directly, not to the outputs of flip-flops or capacitors. ROM of this type is manufactured using a mask during the processing to determine which addresses give 0 and which give 1, so that the complete ROM is referred to as a **masked ROM**. Masked ROM can be cheap if large numbers are being manufactured, but for some purposes this is not always possible. If a new and untried system is to be used, setting up for masked ROM could be a very costly mistake if changes are needed. The alternative is some form of PROM (**programmable read-only memory**), a type of write-once read-often chip. The chips are manufactured unprogrammed, but by connecting to an address-bus and a data-bus, and by using higher voltage supplies than will be used in the computer, the chip can be programmed with a set of 0 and 1 bits. The chip can then be used like a ROM, but with the difference that since the bits are established by the use of a computer, they can readily be changed if a design change is needed. The most popular type of PROM is the EPROM, which allows the pattern of connections to be made electrically by injecting carriers into a semiconductor, and the pattern can be cleared by exposing the material to intense ultraviolet light, a process called **PROM-washing**. If you intend to buy a computer for serious use, as distinct from hobby use, then avoid at all costs any machine whose operating system uses PROMs, since this is almost a guarantee that the makers intend to change the design after the customers start to complain.

Microprocessors

The **microprocessor chip** is the heart of any computer, and is now the universal component of a large number of industrial controllers. The chip consists of a large array of gates and flip-flops, and its unique feature is that internal connections can be made or broken by electrical programming. In other words, a set of binary signals at the data inputs of the microprocessor can be used to set the pattern of gates inside the chip and so determine the action that the chip will carry out with the next set of inputs. It is important to realise that the microprocessor is a serial controller that carries out one action at a time. For example, if the microprocessor is to add two binary numbers, then it must go through a sequence of actions in which at least three inputs will be needed on the data lines. The first action is to place the ADD command in binary form on the data lines. Following this comes the first of the binary numbers to be added. The next item is the second number, and the output from the data lines will then consist of the sum of the numbers. This makes four distinct steps out of a simple addition, and illustrates why the use of a microprocessor can sometimes be too slow for some control actions. The speed at which the steps can be carried out is determined by a clock pulse rate, usually of several MHz.

In a book of this size, we cannot cover all the details of microprocessor action, but a summary will be useful. Newcomers to microprocessor systems are often puzzled by how a sequence of binary numbers can be used in a program in which some numbers are instructions to the microprocessor, some represent numbers or parts of numbers, and some are codes for letters of the alphabet or other characters. The key to the whole process is the sequence of events and the limited actions that the microprocessor can carry out. The actions of the microprocessor consist typically of addition and subtraction, the logic gating actions of AND, OR and NOT, and the ability to shift the bits of a byte around. These, along with reading from and writing to memory addresses, are all the actions that a typical microprocessor IC can do, and the actions of computers are obtained by suitable programming of these actions in sequence.

When a microprocessor has power applied to it, its address lines will take up some starting voltage, typically (though not inevitably) a zero on each line. This will cause one memory location to be read, since the read/write output of the microprocessor will also be set for reading at this time. This memory location must contain the first instruction byte of a program, so that the designer of the system must have something stored into memory at this address. The address will therefore usually be of an EPROM or ROM. Once this first instruction byte has been read, on the first clock pulse, the microprocessor starts a sequence that will continue until it is switched off. The first instruction will be decoded. This may require another instruction to be read from the next address in sequence, or it may require a data byte to be read. Whatever is necessary will be programmed by the gating in the microprocessor, and a counter connected to the address lines will advance the number on the address lines to the next address, one more than the previous one. Here again, suitable data must be placed, either another instruction if that is what is needed, or an item of data such as a number to be added. If, for example, two numbers are to be added, then decoding the first instruction byte will set the gates in the microprocessor in such a way that two more bytes will be read and added. It is the responsibility of the programmer to fill the memory with bytes in the correct sequence and at the correct addresses. Once any instruction has been carried out, the next byte that is read from the next address will be taken as a new instruction, and the sequence repeats. The microprocessor is always carrying out this sequence, and if nothing appears to be happening, it is because the program has forced the microprocessor to keep repeating a set of instructions until some event (such as a signal on a special 'interrupt' pin of the microprocessor) forces it to work with another set of addresses.

Ports

The microprocessor and the memory are two parts of a microprocessor system, and the third essential is a **port**. A port is a means of passing signals to and from circuits that are external to the microprocessor system, the types of circuits that we call **peripherals**. The need for port circuits becomes evident when we look at what is needed in order to pass data into or from a microprocessor circuit. As we have seen, the microprocessor is a programmed chip which operates in a set sequence, with a program stored in the memory containing the steps of instruction for the sequence. Any data that have to be read in by the microprocessor must be connected to the data-bus at a time when the program has reached a step that calls for data to be placed on the data-bus. A port must therefore be able to connect an external set of data lines, an external bus, to the internal bus only at a time when a read from that port is required. The port must therefore be activated by the master clock and also by a signal from the microprocessor, some form of addressing signal. In addition, because the signal that is to be read into the microprocessor system may not exist at precisely the time when it is read, the port needs to contain some memory of its own, so that data can be stored until it can be read. It may be necessary for the port to be able to signal to the microprocessor system that data is stored ready.

The same principles apply to signals that are to be written by the microprocessor system out from the port. These signals have to be placed into the memory of the port, and at the same time, the port may be arranged so that an output signal indicates to the peripheral circuits that data are ready. These data can be read later, and the signal reset so that the microprocessor circuits can then place more data in the memory, because it would clearly be unsatisfactory if the microprocessor system kept storing different data at the port even if it were not being used. The port must therefore provide for these acknowledgement signals, called **handshake** signals, to be passed in each direction.

In addition, the port may have to be able to change the type of signals. One type of signal, the RS-232 serial signal, operates one bit at a time, and with standard voltages of $+12$ V for logic 1 and -12 V for logic 0 (though these are often not adhered to). A port for such inputs and outputs must be able to feed one bit in or out at a time, and also to make the necessary change of logic voltages.

Some ports are general-purpose types, capable of dealing with data inputs or outputs from or to all kinds of peripherals, but particularly keyboards and printers. It is more common now to use specialised types of ports, and for some purposes, notably disc drives, specialised port chips known as **disc drivers** have always been used. The use of specialised chips for purposes such as keyboard connections and video displays greatly reduces the amount of work that the main microprocessor has to carry out, and makes the design of a system much easier. Ports that are intended to convert between parallel transmission and serial transmission (one bit at a time) are known as UART (**universal asynchronous transmitter/receiver**) chips

Servicing

Servicing work on microprocessor circuits demands specialised instruments. For some types of circuits, notably computers, it is possible to run diagnostic programs that will show where faults can be detected, but for smaller-scale equipment, the service engineer has to rely on instruments. The conventional oscilloscope is of very little use except to check that a clock pulse, for example, exists. The specialised instruments range from simple logic probes to very elaborate logic analysers. A **logic probe**, as the name suggests, has a connector which can be put in contact with a bus line and which will indicate the presence of logic 0, logic 1 or pulsing voltages. If the expected voltages on the lines are known, then the logic probe can be used on the microprocessor circuit in much the same way as a DC voltmeter is used in an ordinary audio circuit – to indicate where a fault exists. The advantages of logic probes are that they are inexpensive, easy to use, and are most unlikely to interfere with the circuit action or cause damage. Like the use of the voltmeter in the audio circuit, the intelligent use of a logic probe will solve 90 per cent of the servicing problems in a logic circuit. Those that remain will have to be tackled by the use of instruments such as the logic analyser.

The **logic analyser** is a form of display using an oscilloscope or a printer to show the logic levels on a whole set of bus lines at selected times. The point of this type of display is that it shows a large number of logic levels at one time, so that it is possible to see relationships between levels. The analyser uses its own memory circuits to keep a record of the logic levels on the lines over a number of clock cycles, so that it is possible to show the changes on all the lines in slow motion, looking for incorrect changes that might be the source of problems. Like the oscilloscope, the logic analyser requires some experience to operate and to understand the information that it displays. For complex systems for which there is no guidance on the kind of problems that are likely to be encountered, the use of a logic analyser is indispensable.

15
Practical matters

Passive components such as resistors and capacitors require only two terminals, and there is no need to identify which is which. Transformers will need at least three, usually four, terminals, and these are usually identified by markings on the transformer itself or on paperwork that accompanies the transformer, or on service sheets for the equipment in which the transformer is wired. The main problems of packaging therefore apply to semiconductors, and this has been made particularly complicated by the very large numbers of different connecting arrangements that have been devised in the past. By comparison, digital ICs have almost reached a reasonably standardised situation.

Packages for transistors

Packaging is needed because semiconductors are very small, and the chip itself, whether of a diode, transistor or IC, would be almost impossible to work with and very easily damaged. The chips are therefore mounted into metal or plastic, and thin wire connectors welded to the semiconductor terminals and to the wires or studs that will form the terminals of the package. The chip itself is usually surrounded by some inert material such as silicone jelly or plastic, to prevent contamination from the air. In addition, some means of carrying heat from the semiconductor may need to be used, particularly for power-dissipating devices.

Starting with the simplest types of semiconductors, small diodes are invariably marked at the cathode terminal. The marking may be a red dot or ring, but whatever type is used, the fact that it is on the cathode is sufficient for identification. The type letters or number are more difficult to mark. The very small size of a diode can make these type letterings very difficult to read, particularly when the diode is connected into a circuit, so that a circuit diagram labelled with semiconductor types is particularly useful. Larger diodes for rectification often make use of the 'top-hat' shape of packaging, with the cathode connected to a mounting stud (Figure 15.1). Another common packaging is of four diodes in a bridge connection with the four terminals for a full-wave connection (Figure 15.2). This packaging may incorporate a bolt-hole so that the bridge can be bolted to a metal fin for heat-dissipation.

The packages for transistors can be divided into small-signal types and power types, and similar packages are used for thyristors. The main small-signal packages encountered nowadays are the TO72, TO18, TO5, TO39, TO92, E-line and Silect, and Figure 15.3 indicates the variations that these represent. A considerable complication about such packages as the TO72 is that two transistors using the same packaging may not have identical connections. The only way of finding out the connections is to know the transistor type and look up the connections in a handbook, though for a transistor in a circuit, the components connected to the terminals will often provide useful clues to what the connections are. Determining connections can be particularly difficult when transistors are unmarked. A lot of equipment uses transistors bought in bulk from the manufacturers and marked only with factory internal codes. Unless there are manuals for the equipment that name equivalent transistor types and/or show connections, the servicing of such equipment can be very difficult.

The main principle in packaging is to provide for recognition of leads, and to make the transistor easy to connect into circuit, particularly by automatic means. The case of the transistor is usually made asymmetrical in some way, by a tag, or by flattening one side, so that it should be possible to position the transistor only one way round in automatic insertion machines. Case shapes like the E-line are not always so easy to check, because the difference between the flat side and the rounded side is not easy to see, particularly when the transistor is one of a large number packed on a printed circuit board.

Practical matters 253

Figure 15.1 *Stud-mounting diodes – these are manufactured with a choice of anode or cathode connection to the stud (RS Components Ltd)*

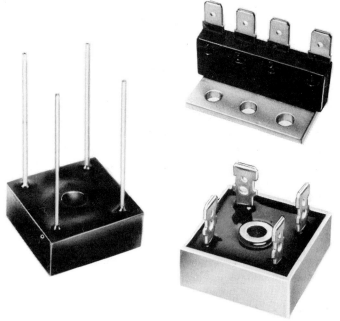

Figure 15.2 *Typical packages of four diodes in a bridge arrangement (RS Components Ltd)*

The packaging styles for the power transistors and thyristors are illustrated in Figure 15.4. The important feature of any packaging for these devices is heat dissipation, and so most of the packages feature metal tabs or studs which are in good thermal contact with the collector, drain or anode of the device. Here

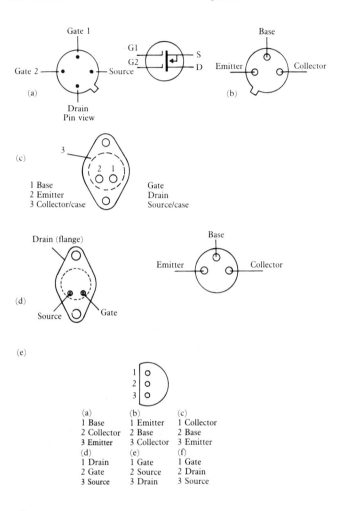

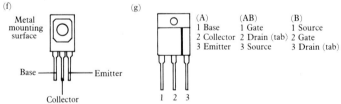

Practical matters 255

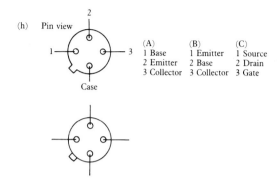

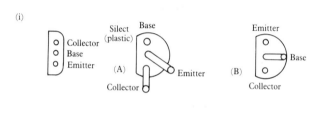

Figure 15.3 *Transistor connections for common configurations (RS Components Ltd)*

Figure 15.4 *Packaging styles for power transistors (RS Components Ltd)*

again, some types of package come in several connection styles. Both thyristors and triacs can also use the type of stud package that is used for some rectifier diodes, but with two terminals in addition to the stud connection (Figure 15.5).

IC Packages

The types of packages used for ICs can show an even greater degree of variation. Some early ICs were packaged like transistors, but with up to nine leads coming out from the can. This type of packaging is still used for some high-frequency amplifier ICs. Later types have tended to use the block style of packaging, with the chip contained in a flat rectangular slab of plastic with the leads on each side of the slab. The most familiar package of this type is the **dual-in-line** (DIL) package, illustrated in Figure 15.6 in its eight-pin form. The pin spacing for this type of pack is standardised so that automatic machinery can be used for drilling printed-circuit boards and so that automatic inserting machines can be used to populate the boards. The DIL principle has been bent slightly for some large chips, such as microprocessors, to

Figure 15.5 *Packaging styles for thyristors and triacs (RS Components Ltd)*

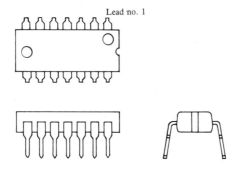

Figure 15.6 *The DIL type of package for ICs. The illustration shows the smaller type of DIL package used for logic chips. A larger version is used for microprocessors and associated chips such as memory and ports*

provide for 40-pin chips with a 0.6in separation between lines. Some more recent microprocessor types have used longer types of DIL packaging in order to accommodate up to 64 pins. A few recent microprocessors have appeared with a square slab that uses a double-decker row of pins along each of the four edges. The names of **pin-grid arrays** or **leadless chip-carriers** have been applied to two styles of package that are already in use (Figure 15.7).

The main aberrations in packaging are found in linear ICs, particularly where power dissipation is involved. A power-

258 Electronics for Electricians and Engineers

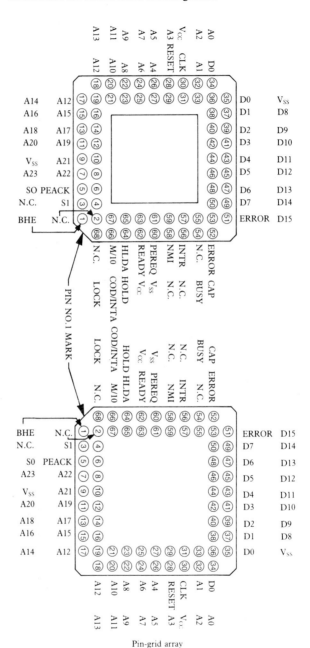

Figure 15.7 *The pin-grid array and leadless chip-carrier packages used for some modern microprocessor types*

Component pad views – As viewed from underside of component when mounted on the board.

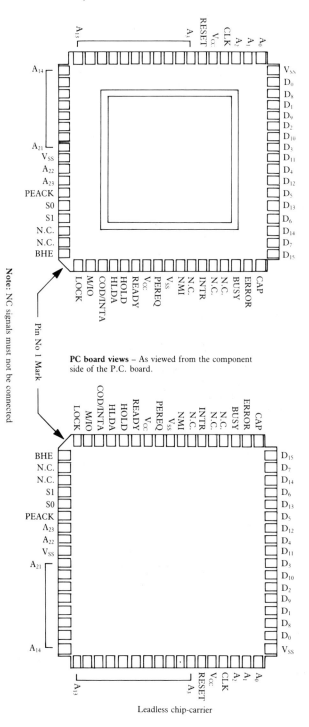

PC board views – As viewed from the component side of the P.C. board.

Leadless chip-carrier

Note: NC signals must not be connected

amplifier IC must provide for several pin connections and also for heat dissipation, and several types incorporate heat-sinks as part of the structure. ICs of this type are very difficult to replace if the original type is no longer available, because though the functions of the chip may be easy to replace, the physical shape and mounting is not, and adaptor boards may be needed. Many of the common types of linear ICs, such as op-amps, are packaged in DIL form, often with eight pins.

Handling

There is an important distinction between transistors and ICs in the sense that transistors are always soldered into circuits, but ICs can be either soldered to a printed-circuit board or plugged into holders that are soldered to the board. The plug-in construction for ICs has been used extensively for small computers, particularly when the design makes it possible to expand the memory of the machine by plugging in more memory ICs. The use of IC holders is not necessarily useful from the point of view of either reliability or servicing, however. From the production point of view, the use of IC holders means that another step is needed, insertion of ICs, before the board is ready for use. The use of insertion tools in production makes the operation relatively easy, but for small scale runs, insertion of ICs without inserting tool can easily cause pin damage, the most common of which is for one pin, which has not been correctly lined up, to hit the side of the holder and be bent under the body of the IC as it is inserted. Visual inspection often fails to spot this, and it is only when pin voltages are read using a logic probe, that the disconnection will be noticed. This mishandling is even more likely when an IC is being replaced, and several manufacturers take the view that it is more satisfactory to solder in all ICs, since the reliability of ICs is such that replacement is unlikely to be needed. There is a lot to be said for this, because holders are by no means completely satisfactory, and as much time can be spent in removing and inserting ICs from and into holders as would be spent in snipping the pins, removing the remains and soldering in a new IC. A further point is that inserting a MOS IC into a holder is more likely to cause electrostatic damage than soldering the IC. If all of the pins are not placed in contact with the holder at the same time, there is a risk that one pin which is isolated could be touched. Though the built-in diodes of MOS ICs will generally protect against damage, the risk is higher than that of soldering.

It is possible to solder in an IC with a wire wrapped around all the pins high up on the shanks. This shorts all of the pins together, preventing any risk of electrostatic damage while the IC pins are being soldered. This is particularly useful when an IC is being replaced during servicing, because the pins will generally be soldered one at a time in such a case. The shorting wire can be removed after all the soldered joints have been checked.

Desoldering of an IC is needed only if there is some doubt about a fault. If the chip is known to be faulty, then desoldering is a waste of time: it is always easier and less harmful to other ICs to snip the pins of the defective IC and then remove them one by one from the board with a hot soldering iron and a pair of pliers. On the few occasions when a chip has to be desoldered and kept in a working state, the use of some sort of desoldering tool is helpful. The type of extended soldering iron bit that covers all the pins of an IC is by far the most satisfactory method, but a separate head is needed for each different size of IC. This technique is particularly useful if the chip is to be tested in a separate tester or in another circuit. It is seldom satisfactory to unsolder an IC and expect to test it in a circuit that makes use of holders, because the presence of even only a film of solder on the pins of an IC make it very difficult to insert into a holder.

If the need to remove an IC by desoldering is rare, then when it does arise it can be tackled by using desoldering braid. This is copper braid which is laid against a soldered joint. When a hot soldering bit is held against the braid, the solder of the joint will melt and will be absorbed by the braid, which can be removed, leaving the joint free of solder. The piece of braid that is not full of solder can then be cut off, and another piece of braid used. This is a slow business, since it has to be repeated for each pin of the IC, but the printed-circuit board is left clean and in good condition. The snag is that a hot iron is being applied to the board many times, and this can cause overheating of other components.

Supplies and signals

Wherever semiconductors are used, great care must be taken over the polarity of supply voltages and signals. Reversal of supply polarity to most ICs, for example, will result in instant destruction, and no form of fusing can avoid this damage. The risk is greatest when several boards are joined to a supply by means of plugs and sockets. The plugs and sockets must be polarised so as to avoid accidental reversal, but even this may not be enough. If an attempt is made, for example, to plug in a supply while the supply is switched on, it is possible that a contact can be made even though the pins of the plug cannot be inserted into the socket. The semiconductors on the circuit board will therefore have been damaged. Even if the power supply has been switched off, the capacitors may be charged, and the attempt to make the incorrect connection will be enough to cause damage. When plugs and sockets are used in this way, power supplies must be switched off before disconnecting, and the power supply capacitors discharged. The power supply should never be switched on while it is disconnected from the other boards, and no attempt should be made to switch on until all plugs have been correctly fitted into their sockets.

The provision of correct signal polarities and amplitudes is equally important, because it is easy to vaporise the base junction of a transistor by connecting it to a large signal pulse. Once again, well-designed equipment will provide very different connectors for inputs and outputs so that signals that are at very different levels will use very different connectors, and some care in assembling the connectors will avoid any problems. The main risk comes during servicing actions that involve signal generators. Servicing of the passive type that makes use only of meters and oscilloscopes involves no risks of this type, other than the possibility of causing shorting by the careless use of probes. When signal generators are used, however, you have to make sure that the polarity and the size of signals will be appropriate to the part of the circuit to which they are applied. As a precaution, signal generators should always be set to provide the minimum possible signal when they are first switched on, and the signal polarity that is needed should always be checked. The important point here is to avoid causing 'service-induced faults'.

One example is connection of a signal generator to the input of an IC, and trying to find the output signal from an output. A common mistake is to connect up, apply a very small signal and to keep increasing the signal amplitude in the hope of seeing an output. It may be that there is a fault at some point between input and output which makes it impossible to obtain an output from the input that is being used, so that increasing the input signal amplitude will simply burn out this input, creating yet another fault in the circuit! A good knowledge of the circuit is the main safeguard here. If you know that an input of 5 mV should be needed at some point, then you should apply a signal of no more than this amplitude. If no output is seen, then instead of increasing the signal amplitude, you should shift the oscilloscope probe to a point nearer the input, until a signal *can* be seen. This enables the true fault to be discovered before another fault is created. For digital equipment, it is much easier to avoid incorrect signal amplitude or polarity, because so many digital systems used the standard +5 V logic signal.

Heat dissipation

Heat dissipation from ICs and transistors is a critical feature of many circuits, and failure to dissipate heat correctly can be the cause of many failures. It is not generally appreciated that the memory boards of computers can run hot, and that by placing one board over another it is possible to reduce the cooling to such an extent as to cause failure. Many of the popular small computers can be fitted with add-on boards that look as if they had been designed by a failed plumber, and it is a tribute to the sturdiness of modern ICs that more faults do not occur. Even the original boards can show evidence that the designers did not spend much time wondering where the power would be dissipated. There is little that can be done if the original design is prone to overheating, other than cut a slit in the box and attach a cooling-fan. Having said that, I should point out that I have tested most of the present generation of small computers for up to ten hours per day continuous running and have never encountered problems with the design, as distinct from add-ons.

The main heat dissipation problems arise when power transistors of ICs are in use, and are attached to a heat-sink. Once again, well-designed original equipment gives very little trouble unless there has been careless assembly of transistors or ICs onto heat-sinks. This can also be a cause of a newly repaired circuit failing again in a very short time. The problem arises because the flow of heat from the collector of a transistor to the metal of a heat-sink is very similar to the flow of current through a circuit. At any point where there is a high resistance to the flow of heat, there will be a 'thermal potential difference' in the form of a large temperature difference. This can mean that the heat-sink body feels pleasantly warm, but the collector junction of the transistor is approaching the danger level. The problem arises because of the connections from one piece of metal to another. Any roughness where two pieces of metal are bolted together will drastically increase the thermal resistance and cause overheating. If either the transistor/IC or the heat sink has any trace of roughness on the mounting surface, or if the surface is buckled in such a way as to reduce the area of contact, then some metalwork with a fine file and emery paper will be needed.

Silicone heat-sink grease should *always* be used on both surfaces before they are bolted together. If the grease is not pressed out from the joint when the joints are tightened, this is an indication that contact is not good. In addition, though, the heat-sink grease is quite a good heat conductor, and will greatly reduce the thermal resistance where two metal surfaces are bolted together. Failure to use heat-sink grease, either at original assembly or later when a transistor or IC is replaced, will almost certainly cause trouble. One of the difficulties about poor heat-sinking is that the problem that it causes may be seasonal, so that the equipment behaves flawlessly throughout the 364-day British winter, only to expire mysteriously on the day of summer.

Index

741 op-amp, 203

AC, 25
AC circuits, 113
A-D converter, 217
Alternator, 94, 113
Ampere, 25
Amplification, 127
Amplitude, 98
Amplitude modulation, 123
AND gate, 223
Annealing, 81
Anode, 105
   cathode ray tube, 136
Anodic protection, 107
Aperiodic amplifier, 185
Aperture grille, 142
Armature, 69, 84
Astable oscillator, 196
Asynchronous counter, 225
Atoms, 1

Balance of bridge, 45
Balanced power supply, 205
Bandwidth, 126, 208
Banking memory, 244
Base, 169

Battery, 28
Bell, Alexander Graham, 186
Biasing bipolar transistor, 181
Bimetallic strip, 58
Binary numbers, 221
Bipolar transistor, 168
Bistable, 199
Bit, 241
Bootstrapping, 188
Breakdown voltage, 150
Bridge circuit, 45
Brushes, 84
Bucket-brigade, 240
Bus lines, 240

Capacitance, 19
Capacitors, 19
Carbon-zinc cell, 27
Carrier storage, 170
Catching diodes, 199
Cathode:
   electrolytic, 105
   thermionic, 137
Cathode rays, 135
CDA, 210
Centre frequency, 189
Chain reaction, 108

Channel, 171
Characteristic, 149
Charge stored in capacitor, 23
Charge, 2
Charging a sphere, 20
Charging capacitor, 129
Chemical effect of current, 30
Chip, 201
Choke, 153
Chopper amplifier, 218
Circuit diagrams, 37
Circuit symbols, 38
Clamping, 156
Class of amplification, 185
Clean machine, 245
Clipping, 156
Clock pulses, 224
CMOS gate, 237
CMOS IC, 231
Coercivity, 78
Collector, 169
Colour CRTs, 142
Colpitts oscillator, 192
Common-base connection, 183
Common-base oscillator, 193
Common-emitter circuit, 182
Commutation, 83, 96
Commutator, 84
Compander, 212
Compound motors, 85
Conducting paths, 29
Conducting solutions, 103
Conductors, 3, 5
Cooling fins, 66
Coulomb's Law, 3
CRO, 121
Curie point, 81
Current-hogging, 170

D-A converter, 218
Damping resistors, 126
Dark current, 160
Darlington circuit, 183
Data bus, 243
DC, 25
DC circuit, 27
DC-restorer, 157
Decibels, 186

Deflection plates, 138
Defluxing coil, 81
Degaussing coil, 81
Delay switches, 60
Delta connection, 101
Demagnetising, 80
Demodulation, 155
Depletion layer, 147
Desoldering ICs, 261
Detached-contact representation, 73
Diamagnetic materials, 74
Dielectric, 19
Differential amplifier, 205
Differentiating circuit, 132
Digital ICs, 220
Digital signals, 128
Digital voltmeter IC, 218
DIL package, 257
Diode bridge, 96
Diodes for commutation, 96
Discharge lighting, 109
Discharging capacitor, 130
Doped crystal, 9
Double-tuned oscillator, 193
Drain, 171
Driver circuits, 134
Dynamic memory, 241

Eddy currents, 100
Effect of voltmeter resistance, **46**
EHT supply, 154
Electric current, 4, 25
Electric field, 14
Electric force, 2
Electric motor, 82
Electricity distribution, 63
Electrode, 104
Electrolyte, 104
Electrolytic capacitor, 23
Electrolytic cell, 104
Electrolytic corrosion, 106
Electrolytic timers, 105
Electromagnetic induction, **94**
Electromagnetic wave, 18
Electron gun, 138
Electrons, 1
Electroplating, 105

Index

Electrostatic air cleaners, 15
Electrostatic force, 2
Electrostatic loudspeakers, 17
Electrostatic precipitators, 16
Electrostatic voltmeters, 14
Electrostatics, 3
Elements, 1
EMF, 26
Emitter, 169
Emitter-follower, 182
Enable pins, 244
Enhancement, 172
Equivalent circuits, 47
Expander, 231

Fanin, 228
Fanout, 228
Farad, 19
Faraday, Michael, 82
FET, 170
    switching circuits, 176
Field magnet, 82
FIFO memory, 241
Filament, 136
Fleming's LH rule, 93
Fleming's RH rule, 94
Flip-flop, 199, 224
Fluorescent lighting, 109
Flux density, 74
Flux leakage, 100
Flux lines, 74
Fluxmeters, 93
Flyback, 139
Force effects of current, 69
Forward bias, 148
Forward current gain, 169
Frequency, 96, 113
FSD, 88
Fundamental, 120
Fuses, 62

Gain, 126
Ganged tuning, 128
Gate, 171
Gates, 222
Generator principle, 95

Hall, 9

Hall effect, 92
Handling CMOS, 237
Handshake signals, 249
Hard magnetic material, 69
Harmonics, 120
Hartley oscillator, 192
HC/HCT CMOS, 239
Heat dissipation, 36, 263
Heat sink, 64
Heating effect of current, 20, 56
Heating elements, 57
Hertz, Heinrich, 18
Hertz, unit, 98
Holes, 6
Hysteresis loop, 76

ICs, 202
Impedance, 119
Impurity effects, 6
Indirectly-heated cathode, 137
Insulators, 3
Integrating circuit, 133
Integration, 201
Intermediate frequency, 129
Interrupt pin, 248
Inverting amplifier, 205
Ionised gas, 107
Ions, 4, 103

JFETs, 176
JK flip-flop, 224
Joule effect (heating), 29
Junctions, 146

Kirchhoff's Laws, 29

Lag, 117
Latching relay, 72
LCD, 163
LDR, 161
Lead, 117
Leadless chip carrier, 258
LED, 161
Local oscillator, 127
Logic analyser, 250
Logic probe, 250
LS-TTL, 230

Magnetic effect of current, 29
Magnetic field, 67
Magnetic hysteresis, 76
Magnetic materials, 73
Magnetic shields, 80
Magnetism, 4
Magnetomotive force, 75
Masked ROM, 246
Maxwell, Clark, 18
Mean free path, 136
Measuring current, 30
Memory, 241
   organisation, 243
Meter scales, 88
Meter shunts, 89
Microprocessor, 246
Modulation, 123
Monostable, 198
MOSFET, 171
Multilayer diodes, 165
Multiplier, 153
Multivibrator, 196
Mumetal shield, 81
Mutual conductance, 173

N-type, 9, 146
NAND gate, 223
Negative feedback, 180
Neon lighting, 109
Noise immunity, 229
Non-inverting amplifier, 207
Non-linear distortion, 182
Non-linear resistance, 54
Non-volatile memory, 241
NOR gate, 223
Norton amplifier, 210
Norton's theorem, 55
NOT gate, 223
Nucleus, 1

Ohm's Law, 32
Ohmic circuit, 32
One-shot, 198
Op-amps, 203
Opto-isolators, 164
Optoelectronics, 159
OR gate, 223
Oscillator, 191

Oscillator pulling, 173
Oscilloscope, 121, 139
Overheating, 36

P-type, 9, 146
Packaging, transistor, 251
Parallel components, 37
Parallel connection, cells, 29
Parallel resonant circuit, 125
Parallel-plate capacitor, 20
Paramagnetic materials, 74
PD (potential difference), 12
Peak amplitude, 98, 113
Peak-to-peak value, 98
Peripherals, 248
Phase, 100
Phase control of thyristor, 215
Phase relationships, 117
Phase shift, 115
Phase-locked loop, 155
Phosphor stripes, 143
Photodiodes, 159
Photovoltaic cells, 161
Pin-grid array, 258
Pinchoff, MOSFET, 172
Ports, 248
Positive feedback, 127
Post-deflection acceleration, 140
Potential, 12
Potential divider, 42
Potentiometer, 43
Power, 57
Power amplifiers, AF, 211
Power MOSFETs, 174
Power output, 187
PPI, 144
Preamplifiers, AF, 211
Preferred values, 40
Preferred voltage series, 158
Preset, 43
Primary cell, 28
Primary coil, 99
PROM, 246
Propagation time, 227
Pulse, 25

Q-factor, 194
Quartz crystal, 194

Index

Radar CRTs, 144
Radial field, 86
Radio waves, 122
RAM, 242
Ramp, 139
Random access, 242
Reactance, 118
Rectification, 151
Reed relay, 72
Refreshing, 242
Relative permeability, 75
Relative permittivity, 21
Relaxation oscillator, 192
Relay switching with transistors, 178
Relay, 71
Remanence, 78
Reservoir capacitor, 151
Resistance, 32
Resistance and resistivity, 7
Resistance measurement, 91
Resistivity, 6, 33
Resonant frequency, 124
Reverse bias, 148
Ripple, 151
RMS, 114
Rolled capacitor, 22
ROM, 246
Rotor, 96
RS-232, 249

Sacrificial anode, 107
Salts, 103
Saturation, 100
Saturation flux density, 77
SAW filter, 191
Schottky diode, 150
Schottky TTL, 230
SCS, 167
Secondary breakdown, 170
Secondary cell, 28
Secondary coil, 99
Selection, 125
Semiconductor diode, 149
Semiconductors, 6
Sequential circuits, 224
Series components, 37
Series connection, cells, 28

Servicing microprocessors, 250
Seven-segment display, 163
Shaded-pole motors, 86
Shunt field, 85
Signal generators, 122
Single-ended push-pull, 189
Sinking current, 227
Slew rate, 208
Slip-rings, 96
Soft magnetic material, 68
Solenoid, 67
Source, 171
Sourcing current, 228
Stagger tuning, 126
Star connection, 101
State table, 224
Static memory, 241
Stator, 96
Stray capacitance, 21, 195
Superhet, 126
Superposition, 51
Supplies and signals, 261
Switch-mode supply, 216
Switch start, fluorescent light, 110
Switching, 194
Synchronous motors, 86

Television CRTs, 141
Temperature and resistivity, 6
Temperature coefficient of resistance, 33
Thermal switches, 58
Thermionic emission, 135
Thermistor, 61
Thermocouple, 62
Thevenin's theorem, 48
Thin-film resistors, 34
Three-phase supplies, 100
Thyratron, 110
Thyristors, 165
Time constant, 130
Toggle, 199
Torque, 85
Totem-pole circuit, 189
Transformer laws, 99
Transformer start, fluorescent light, 110
Transformers, 98

Transistors, 168
Triac, 166
Trigger pulse, 198
Tropicalised components, 106
Truth tables, 223
TTL, 226
Tuned amplifier, 189
Tuning, 123

UART chips, 249
Unbalanced power supply, 206
Units, electrical, 35
Universal shunt, 90
Untuned amplifier, 185

Varicap diodes, 158
Video ICs, 213
VMOS, 175
VOGAD, 211
Volatile, 241
Volt, 12
Voltage amplification, 178
Voltage regulation, 214
Voltage step, 131

XOR gate, 223

Zener diode, 150, 157
Zero-voltage switching of thyristor, 215